Progress in Molecular and Subcellular Biology

Series Editors   W. E. G. Müller (Managing Editor)
Ph. Jeanteur, Y. Kuchino, M. Reis Custódio,
R.E. Rhoads, Đ. Ugarković

50

Robert E. Rhoads

Editor

miRNA Regulation of the Translational Machinery

 Springer

Editor
Robert E. Rhoads
Louisiana State University
Health Sciences Center in Shreveport
Department of Biochemistry
and Molecular Biology
1501 Kings Highway
Shreveport LA 71130-3932
USA
rrhoad@lsuhsc.edu

ISSN 0079-6484
ISBN 978-3-642-26140-4   e-ISBN 978-3-642-03103-8
DOI 10.1007/978-3-642-03103-8
Springer Heidelberg Dordrecht London New York

Cover design: Boekhorst Design BV, The Netherlands

Printed on acid-free paper

Springer is part of Springer Science+Business Media (www.springer.com)

Preface

An odd and unexpected finding was reported by the laboratory of Richard Jorgensen in 1990: expression of extra copies of the gene encoding chalone synthase in petunias turned off the endogenous chalone synthase gene. An observation that appeared totally unrelated was made by the laboratory of Victor Ambrose in 1993: a gene in *Caenorhabditis elegans*, *lin-4*, controlled the timing of larval development but did not encode a protein. Rather, it expressed two small RNAs that were complementary to the 3'-untranslated region of the *lin-14* gene in a region that had previously been shown to repress expression of the LIN-14 protein. From another quarter, David Baulcombe's laboratory showed in 1997 that plant viruses could induce sequence-specific gene silencing. Then in a landmark paper, Andrew Fire and Craig Mello showed in 1998 that double-stranded RNA (dsRNA) triggers a gene-silencing mechanism that they dubbed RNA interference (RNAi), for which discovery they were awarded the Nobel Prize in Physiology or Medicine in 2006.

These diverse findings have triggered an explosion of research around the world in both plants and animals to discover the mechanisms and broader ramifications of RNAi. We now know that there are both exogenous pathways involving formation of siRNA when dsRNA is introduced and endogenous pathways involving miRNA, piwiRNA, and rasiRNAs. All pathways culminate in formation of an RNA-induced silencing complex (RISC) containing a member of the Argonaute protein family bound to a 22-nt RNA strand that interacts with a target mRNA or gene through Watson-Crick base pairing.

The predominant mechanism of gene silencing involves RISCs containing miRNAs. David Bartel has estimated that as much as one-third of all human genes are regulated by miRNAs. Initial studies focused on mechanisms involving either formation of heterochromatin or mRNA degradation, but then articles began to appear showing that RISCs can also interfere with the translational machinery. Published reports indicate that translation can be inhibited through various mechanisms – binding of Argonaute to the mRNA cap to prevent its interaction with eIF4E, inhibition of 80S initiation complex formation by interference with eIF6, and inhibition of translational elongation. Recent reports also show that some miRNAs can enhance translation. Yet, there is not universal agreement in the field of how miRNAs affect the translational machinery.

The chapters collected in this volume represent contribution by leaders in the search to understand how miRNAs affect translation. They include chapters representing work in plants and *C. elegans*, the biological systems that originally led to the discovery of RNAi, and also include chapters on mammalian systems, with special emphasis on regulation of a key tumor suppressor and a protein that restricts human immunodeficiency virus 1 (HIV-1).

Regulation of gene expression by miRNAs plays critical roles in malignant transformation and development of cardiovascular disease. There is currently intense activity to develop miRNAs as therapeutic agents to combat such diseases. Yet, incomplete knowledge of how miRNAs accomplish gene silencing hinders progress in this area. The authors of this volume are making important contributions toward understanding this phenomenon.

Robert E. Rhoads
Shreveport, Louisiana, USA
September 2009

Contents

Contributors

Traude H. Beilharz
Molecular Genetics Division, Victor Chang Cardiac Research Institute (VCCRI),
Darlinghurst (Sydney), NSW 2010, Australia; School of Biotechnology &
Biomolecular Sciences and St Vincent's Clinical School, University of New South
Wales, Sydney, NSW 2052, Australia

Xavier C. Ding
Friedrich Miescher Institute for Biomedical Research (FMI), Maulbeerstrasse 66,
WRO-1066.1.38, CH-4002 Basel, Switzerland

Marc R. Fabian
Department of Biochemistry, McGill University, Montréal, QC H3G 1A1,
Canada; Goodman Cancer Center, McGill University, Montréal,
QC H3G 1A1, Canada
marc.fabian@@mcgill.ca

Fátima Gebauer
Centre de Regulació Genòmica (CRG-UPF), Gene Regulation Programme,
Dr Aiguader 88, 08003 Barcelona, Spain

Helge Großhans
Friedrich Miescher Institute for Biomedical Research (FMI), Maulbeerstrasse 66,
WRO-1066.1.38, CH-4002 Basel, Switzerland
helge.grosshans@fmi.ch

David T. Humphreys
Molecular Genetics Division, Victor Chang Cardiac Research Institute (VCCRI),
Darlinghurst (Sydney), NSW 2010, Australia

Benjamin A. Hurschler
Friedrich Miescher Institute for Biomedical Research (FMI), Maulbeerstrasse 66,
WRO-1066.1.38, CH-4002 Basel, Switzerland

Aida Martínez-Sánchez
Centre de Regulació Genòmica (CRG-UPF), Gene Regulation Programme,
Dr Aiguader 88, 08003 Barcelona, Spain

Thomas Preiss
Molecular Genetics Division, Victor Chang Cardiac Research Institute (VCCRI),
Darlinghurst (Sydney), NSW 2010, Australia; School of Biotechnology &
Biomolecular Sciences and St Vincent's Clinical School, University of New South
Wales, Sydney, NSW 2052, Australia
t.preiss@victorchang.edu.au

Nahum Sonenberg
Department of Biochemistry, McGill University, Montréal, QC H3G 1A1,
Canada; Goodman Cancer Center, McGill University, Montréal, QC H3G 1A1,
Canada, Nahum
Sonenberg@mcgill.ca

Thomas R. Sundermeier
Department of Biochemistry, McGill University, Montréal, QC H3G 1A1,
Canada; Goodman Cancer Center, McGill University, Montréal,
QC H3G 1A1, Canada
Thomas.Sundermier@mcgill.ca

Motoaki Wakiyama
RIKEN, Systems and Structural Biology Center, 1-7-22 Suehiro-cho,
Tsurumi, Yokohama 230-0045, Japan
waki@gsc.riken.jp

Hai Wang
School of Biological Science & Center for Plant Science Innovation,
University of Nebraska, Lincoln, Lincoln, NE 68588, USA

Shigeyuki Yokoyama
RIKEN, Systems and Structural Biology Center, 1-7-22 Suehiro-cho,
Tsurumi, Yokohama 230-0045, Japan; The University of Tokyo,
Graduate School of Science, 7-3-1 Hongo, Bunkyo, Tokyo 113-0033, Japan
yokoyama@biochem.s.u-tokyo.ac.jp

Bin Yu
School of Biological Science & Center for Plant Science Innovation,
University of Nebraska, Lincoln, Lincoln, NE 68588, USA
byu3@unl.edu

Hui Zhang
Center for Human Virology, Division of Infectious Diseases, Department of
Medicine, Thomas Jefferson University, Philadelphia, PA 19107, USA
hui.zhang@jefferson.edu

Chapter 1
Understanding How miRNAs Post-Transcriptionally Regulate Gene Expression

Marc R. Fabian, Thomas R. Sundermeier, and Nahum Sonenberg

Contents

Abstract The discovery of microRNA (miRNA)-mediated gene silencing has added a new level of complexity to our understanding of post-transcriptional control of gene expression. Considering the ubiquity of miRNA-mediated repression throughout basic cellular processes, understanding its mechanism of action is paramount to obtain a clear picture of the regulation of gene expression in biological systems. Although many miRNAs and their targets have been identified, a detailed understanding of miRNA action remains elusive. miRNAs regulate gene expression at the post-transcriptional level, through both translational inhibition and mRNA destabilization. Recent reports suggest that many miRNA effects are mediated through proteins of the GW182 family. This chapter focuses on the multiple and potentially overlapping mechanisms that miRNAs utilize to regulate gene expression in eukaryotes.

M.R. Fabian, T.R. Sundermeier, and N. Sonenberg (✉)
Department of Biochemistry, McGill University, Montréal, QC H3G 1A1, Canada;
Goodman Cancer Center, McGill University, Montréal, QC H3G 1A1, Canada
e-mail: marc.fabian@mcgill.ca, Thomas.Sundermeier@mcgill.ca and
Nahum.Sonenberg@mcgill.ca

R.E. Rhoads (ed.), *miRNA Regulation of the Translational Machinery*,
Progress in Molecular and Subcellular Biology 50,
DOI 10.1007/978-3-642-03103-8_1, © Springer-Verlag Berlin Heidelberg 2010

1.1 Introduction

MicroRNAs (miRNAs) are small RNA molecules, approximately 22 nucleotides in length, encoded within the genomes of more eukaryotes. miRNAs direct an intricate mechanism that regulates eukaryotic gene expression at the post-transcriptional level. miRNA functions in mammals include modulating hematopoietic lineage differentiation, insulin secretion, apoptosis, heart muscle development, neuron development, and many other processes (Chen and Lodish 2005; Poy et al. 2004; Welch et al. 2007). Furthermore, control of gene expression through miRNA activity has been shown to play a significant role in numerous human pathologies, including cancer (Calin and Croce 2006a, 2006b; Chang and Mendell 2007; Croce and Calin 2005; Cummins and Velculescu 2006; Dalmay and Edwards 2006; Garzon et al. 2006; Giannakakis et al. 2007; Hammond 2006; He et al. 2007; Mattes et al. 2008; Stefani 2007). Recent research has shed some light on the mechanisms by which miRNAs regulate gene expression; however, many studies have yielded contradictory conclusions. Overall, miRNAs regulate gene expression by inhibiting mRNA translation and/or facilitating mRNA degradation.

1.1.1 Eukaryotic Translation

Translation may be divided into three steps: initiation, elongation, and termination. Initiation involves the assembly of an 80S ribosome complex positioned at the appropriate start site on the mRNA to be translated. Elongation is the polypeptide synthesis step, where the nucleotide sequence carried on the mRNA molecule is translated into the amino acid sequence of the growing peptide chain. Termination involves the release of the newly synthesized protein. In eukaryotes, the rate-limiting step under most circumstances is initiation. Consequently, initiation is the most common target for translational control. All nuclear transcribed eukaryotic mRNAs contain at their 5′ end the structure m⁷GpppN (where N is any nucleotide) termed the "cap," which facilitates ribosome recruitment to the mRNA. This canonical mechanism of translation initiation is termed as cap-dependent translation initiation. In contrast, many eukaryotic and viral mRNAs are translated via alternative, cap-independent, mechanisms.

1.1.1.1 Cap-Dependent Translation Initiation

Cap-dependent translation depends on the activities of a variety of eukaryotic initiation factors (eIFs). It is accomplished through mRNA scanning mechanism, whereby the small (40S) ribosomal subunit, in complex with a number of eIFs, binds the mRNA near the 5′ cap structure and scans the mRNA in a 5′–3′ direction until it encounters an AUG start codon in an optimal context (Kozak 1978; Kozak and Shatkin 1979) (Fig. 1.1a). Recruitment of ribosomes to a given mRNA is

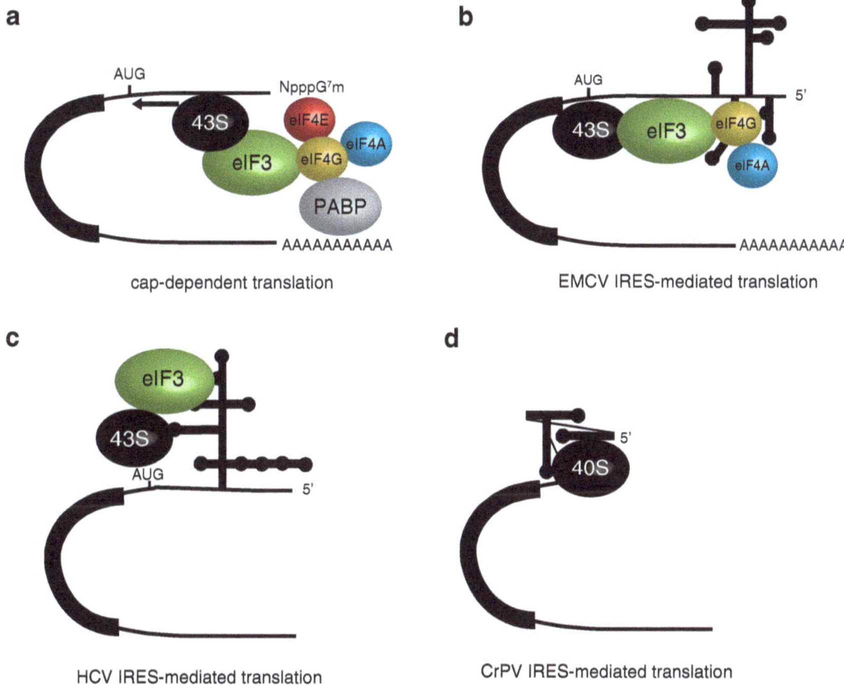

Fig. 1.1 Mechanisms of translation initiation. (**a**) cap-dependent translation. (**b**) EMCV IRES-mediated cap-independent translation. (**c**) HCV IRES-mediated cap-independent translation. (**d**) CrPV intergenic IRES-mediated cap-independent translation. Open reading frames are denoted as *thick curved black lines*. IRES secondary structures are presented as *thick black lines* bound by translation factors and/or ribosomal subunits

facilitated by the 5′ cap, the 3′ poly(A) tail, the poly(A) binding protein (PABP), and the eIF4F complex. eIF4F is a three subunit complex (Edery et al. 1983; Grifo et al. 1983) composed of (1) eIF4A, an ATP-dependent RNA helicase that unwinds secondary structures, (2) eIF4E, a 24 kDa polypeptide that specifically interacts with the cap structure (Sonenberg et al. 1979), and (3) eIF4G, a large scaffolding protein that binds to both eIF4E and eIF4A. The poly(A) tail functions as a translational enhancer (Sachs 2000), as the 3′ poly(A) and 5′ cap structure act synergistically to enhance translation initiation (Gallie 1991; Sachs and Varani 2000). This synergy can be explained by the physical interaction between PABP and eIF4G that brings about the circularization of the mRNA. mRNA circularization is thought to increase the affinity of eIF4E for the cap, thus enhancing the rate of translation initiation. A given mRNA is activated when the eIF4F complex binds to the 5′ cap (through eIF4E) and interacts with the 3′ poly(A) tail (through eIF4G–PABP–poly(A) interaction). The activated mRNA is then bound by the 43S preinitiation complex (PIC), which contains the 40S ribosomal subunit, an initiator tRNA (Met-tRNAi), as well as eIFs 1, 1A, 2, 3, and 5. eIF3 is a large multisubunit scaffolding protein that bridges the mRNA complex to the ribosome through interaction with eIF4G (Imataka et al. 1997).

Once the PIC associates with the mRNA, it proceeds to scan the 5′ untranslated region (UTR) until an appropriate initiation codon is encountered. The RNA helicase activity of eIF4A is thought to promote scanning. eIF4B functions, at least in part, to enhance the RNA helicase activity of eIF4A, likely by increasing the affinity of eIF4A for ATP (Bi et al. 2000; Rogers et al. 1999). eIF1 and eIF1A are thought to promote scanning and enhance the fidelity of start codon selection (Pestova and Kolupaeva 2002), while eIF5 is the GTPase-activating protein for eIF2 (Mitchell and Lorsch 2008). Once the start codon is recognized, the partially hydrolyzed phosphate from the eIF2-bound GTP is released; this is then followed by release of eIF1 (Algire et al. 2005; Maag et al. 2005). The 60S ribosomal subunit then joins the 43S initiation complex, with the assistance of eIF5B, and translation elongation commences (Pestova et al. 2000). eIF6 is the only initiation factor currently known to regulate the availability of the 60S subunit (Ceci et al. 2003). Free 60S ribosomal subunits bound by eIF6 are unable to bind to the 40S subunit to form 80S ribosome complexes. Only when an eIF6-bound 60S is phosphorylated by RACK1/PKC, it can dissociate from the 60S and allow it to join the 40S subunit upon start codon recognition.

1.1.1.2 Poly(A) Tail-Independent Translation Initiation

The poly(A) tail plays a critical role in the control of translation initiation under many physiological conditions (Wickens et al. 2000). Histone mRNAs are the only mammalian mRNAs that lack poly(A) tails; nevertheless, they are efficiently translated. A terminal stem-loop on histone mRNAs binds the histone stem-loop binding protein, which functionally substitutes for PABP by interacting with eIF4G (Ling et al. 2002).

1.1.1.3 Cap-Independent Translation Initiation

The discovery, in picornaviruses two decades ago (Jang et al. 1988; Pelletier and Sonenberg 1988), of internal ribosome entry sites (IRESes) has added a new degree of complexity to our understanding of translation initiation. IRESes are generally (but not always (Gilbert et al. 2007)) highly structured *cis*-acting RNA elements that function to enhance translation initiation in a cap-independent manner. Although originally discovered in viral genomes, IRESes have since been found in several mRNAs (i.e., myc, XIAP, and DAP5 (Henis-Korenblit et al. 2000; Holcik et al. 1999; Stoneley et al. 1998)). Often, but not always (as in the case of the Cricket paralysis virus (CrPV) intergenic IRES (Wilson et al. 2000)) located in the 5′UTR, IRESes enhance translation in the absence of eIF4E by recruiting the 40S subunit to the mRNA through unconventional means. Certain IRESes (such as those of poliovirus and encephalomyocarditis virus (EMCV)) (Fig. 1.1b) can directly bind the eIF4G subunit of the eIF4F complex, thus bypassing the requirement for eIF4E and the 5′ cap (Hellen and Wimmer 1995; Kolupaeva et al. 1998). The Hepatitis C virus IRES bypasses the need for the entire eIF4F complex and binds directly to eIF3 and the 40S ribosomal subunit (Pisarev et al. 2005) (Fig. 1.1c). The CrPV intergenic IRES enhances translation via a factorless mechanism, whereby the IRES mimics an aminioacylated tRNA and positions itself within the P-site of the ribosome (Jan and Sarnow 2002; Spahn et al. 2004). This allows the CrPV intergenic IRES to initiate translation from a non-AUG codon (Fig. 1.1d).

1.2   miRNA Biogenesis

miRNAs are small (~22 nucleotides) noncoding RNAs that were first discovered in *C. elegans* but have since been found to exist in almost all eukaryotes ranging from plants to insects to mammals (Lee and Ambros 2001) (*S. cerevisiae* is an exception). More recently, miRNAs have also been found within the genomes of several viruses including Epstein–Barr virus and several herpesviruses (Cullen 2009). miRNAs are processed from primary transcripts via a two-step mechanism involving two RNase III-type enzymes known as Drosha and Dicer (Fig. 1.2). miRNAs are transcribed, either from discreet miRNA genes or as parts of introns of protein coding genes. These initial miRNA precursors, known as pri-miRNAs, are processed into ~70 nt hairpin structures known as pre-miRNAs by a nuclear enzyme complex known as the microprocessor. The microprocessor contains an endoribonuclease known as Drosha as well as a double-stranded RNA binding protein known as DiGeorge syndrome critical region 8 (DGCR8) in mammals and Partner of Drosha (Pasha) in *D. melanogaster* and *C. elegans*. DGCR8/Pasha is required for proper pri-miRNA processing (Han et al. 2006; Zeng and Cullen 2003, 2005). Drosha contains two RNase domains that cleave the 5′ and 3′ ends, releasing the pre-miRNA (Han et al. 2004). By binding specifically to the pri-miRNA dsRNA hairpin, DGCR8/Pasha determines the cleavage sites on the pri-miRNA, and hence the length of the pre-miRNA (Han et al. 2006). In some instances, the sequence of the mature pre-miRNA corresponds precisely to the sequence of a spliced intron. These spliced-out pre-miRNAs, known as mirtrons, no longer require microprocessor activity in order to generate mature miRNAs (Berezikov et al. 2007; Okamura et al. 2007; Ruby et al. 2007). In the case that pre-miRNAs are present in introns, recent results question the temporal order of pre-mRNA splicing and miRNA processing. Drosha-dependent pre-miRNA processing can still occur on pre-mRNAs that are splicing-deficient, suggesting that Drosha can process intronic pre-miRNAs directly from pre-mRNAs (Kim and Kim 2007). A more recent study, using an in vitro system displaying both splicing and pre-miRNA processing, demonstrated that microprocessor-dependent pre-miRNA cropping can occur kinetically faster than splicing (Kataoka et al. 2009). This study concluded that the microprocessor and spliceosome may be functionally linked such that Drosha-mediated miRNA processing and pre-mRNA

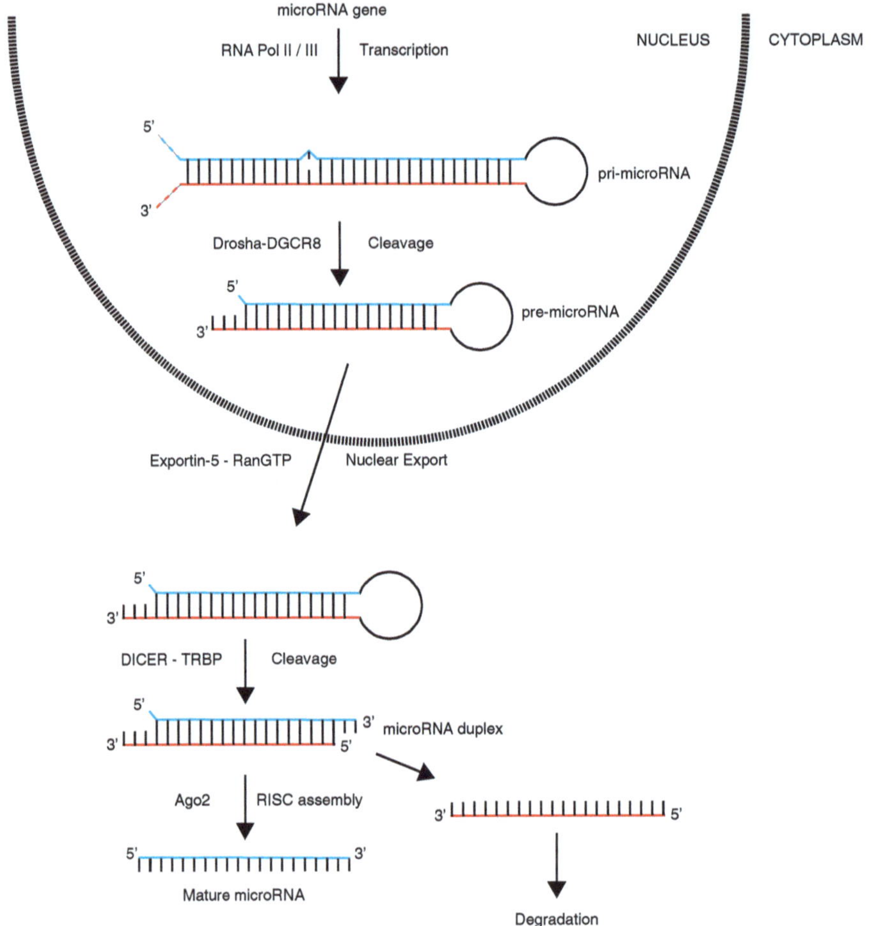

Fig. 1.2 Mechanism of microRNA biogenesis. The nuclear membrane is presented as a *broken curved black line*

splicing may occur simultaneously. Pre-miRNAs are transported from the nucleus to the cytoplasm (by a complex of Exportin5 and Ran-GTP (Yi et al. 2003)), where they are processed into ~22 bp double stranded RNAs by the RISC loading complex. The RISC loading complex consists of the RNase Dicer, the dsRNA binding protein TRBP (product of the *loquacious* gene in flies), PACT (protein activator of PKR), and Argonaute proteins. While the number of Dicer paralogues varies evolutionarily, vertebrates have only one gene coding for a Dicer-like protein. Once processed, one of two strands of the miRNA is loaded into a ribonucleoprotein complex, referred to as a miRNA-induced silencing complex (miRISC). The most widely studied protein components of miRISCs are proteins of the Argonaute family.

Pre-miRNA processing by Dicer and miRISC complex assembly (loading of the mature miRNA onto Argonaute proteins) are thought to occur simultaneously at the RISC loading complex.

1.3 miRNA-Mediated Regulation of Eukaryotic Gene Expression

In most cases, miRNA-targeted sites are located in mRNA 3′UTRs. miRNAs can also regulate gene expression of mRNAs that contain miRNA target sites in their 5′UTR (Lytle et al. 2007); however, there is currently only one known example of a miRNA targeting the 5′UTR of naturally occurring mRNA (Orom et al. 2008). A recent report has shed some light on the nature of the evolutionary preference for miRNA target sites to reside in the mRNA 3′UTR. Gu et al. reported that, mutating the stop codon of reporter mRNAs such that the coding sequence extends past miRNA target sites, thus positioning the target sites within the mRNA coding sequence, significantly impairs miRNA-dependent repression of reporter mRNA translation (Gu et al. 2009). However, placing rare codons upstream of target sites within the coding sequence partially restored miRNA-mediated repression. These results suggest that actively translating ribosomes may displace the miRISC complex from target sites positioned within the coding sequence. Nevertheless, experimentally validated miRNA target sites have been reported in the coding sequences of several genes (Forman et al. 2008; Rigoutsos 2009). One interesting example is the presence of three let-7 target sites within the coding sequence of *dicer*, which could represent a negative feedback loop for production of the mature form of this miRNA.

Specificity of miRNA function is controlled through the direct base pairing of a miRNA-loaded RISC to miRNA-complementary target sites on targeted mRNAs (Doench and Sharp 2004). miRNA-regulated mRNAs often harbor multiple miRNA target sites within their 3′UTRs, sites that in many cases are phylogenetically conserved between species (Stark et al. 2005). miRNAs are roughly the same size as small-interfering RNAs (siRNAs) but are not generated and do not act for the most part in the same manner. Although both miRNAs and siRNAs interact with Argonaute (Ago) proteins, miRNAs are distinct from siRNAs in that, unlike siRNAs, miRNAs imperfectly base pair to target sites and do not lead to endonucleolytic cleavage of targeted mRNAs, but rather regulate their expression by other means (Bartel 2004). Interestingly, siRNAs can act as miRNAs if made to base pair imperfectly to target sites (Zeng et al. 2003), and miRNAs can act as siRNAs if made to base pair perfectly (Doench et al. 2003).

Considering the short length of time that has past since their discovery, a wealth of effort and resources have been expended in an attempt to elucidate exactly how miRNAs mediate their effects. However, the mechanism by which miRNAs exert post-transcriptional control of gene expression remains highly controversial. Early reports generally suggested that miRNAs inhibit gene expression at the post-transcriptional level, at some stage post-translation initiation. These reports also suggested that miRNA action had little or no effect on the abundance or stability of target mRNAs. More recent results challenge these data as results from both in vitro and in vivo studies have shown that miRNAs can inhibit translation initiation as well as promote decay of target mRNAs. As such, the literature now contains reports favoring three different potential modes of miRNA-mediated repression: miRNAs may (1) destabilize target mRNAs, (2) inhibit translation initiation, or (3) block translation at some stage after initiation (Fig. 1.3). These three possible inhibitory mechanisms are by no means mutually exclusive. It is possible that the primary mode of miRNA mediated gene regulation may vary by cell type or developmental stage, possibly controlled by miRNA levels or miRISC complex components. Perhaps the most compelling evidence for cellular regulation of the nature of the miRNA response comes from recent reports suggesting that serum starvation can switch the miRNA response from inhibition of target gene expression to enhancement (Vasudevan et al. 2007, 2008). Indeed, the nature of miRNA control of gene expression is much more complex than initially thought.

1.3.1 miRNA-Mediated Translational Control

miRNAs are studied in a variety of in vivo and in vitro systems derived from mammals, flies, and worms. miRNAs first made their grand entrance in the study of developmental timing in *C. elegans*. Genetic analyses carried out by Victor Ambros'

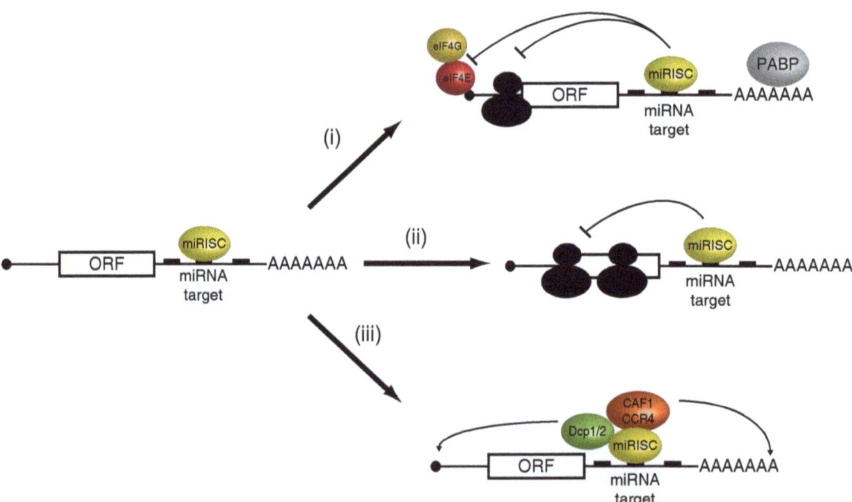

Fig. 1.3 Mechanisms of miRNA-mediated repression. (1) inhibition of translation initiation. (2) inhibition of polysome elongation. (3) miRNA-mediated deadenylation and decapping via the CAF1/CCR4 deadenylase complex and Dcp1/2 decapping complex

and Gary Ruvkun's laboratories determined that the *lin-4* gene functioned to repress the production of *lin-14* protein. The key discoveries were that the *lin-4* gene did not encode a protein but rather a short noncoding RNA (what is now called a miRNA), and that the *lin-4* miRNA exhibited partial complementarity to conserved sequences within the 3'UTR of *lin-14* mRNA, sites that were important for developmental regulation (Lee et al. 1993; Wightman et al. 1991, 1993). It was later demonstrated that lin-4 repressed lin-14 protein production at the translational level with no observed effect on lin-14 mRNA transcription or stability (Olsen and Ambros 1999). The possibility that miRNAs represented a general phenomenon, rather than a species-specific one, came with the discovery of a second miRNA in *C. elegans*, let-7 (Pasquinelli et al. 2000; Reinhart et al. 2000). Just like the lin-14 miRNA, the let-7 miRNA also regulated the expression of a target mRNA (*lin-41*); however, unlike the lin-14 miRNA, the let-7 miRNA was phylogenetically conserved in both flies and animals. Data for both the lin-4 and let-7 miRNAs suggested that they did not influence mRNA biogenesis or stability (although mRNA degradation of let-7 and lin-4 targeted mRNAs has since been reported in *C. elegans* (Bagga et al. 2005; Ding and Grosshans 2009)), but rather inhibited translation. Polyribosome sedimentation experiments conducted by the Ambros' laboratory determined that both lin-4 miRNA and its targeted *lin-14* mRNA were still associated with polyribosomes, suggesting that the lin-4 miRNA inhibits translation at a post-initiation step (Olsen and Ambros 1999). An identical distribution of polyribosomes was described for the lin-4 miRNA-repressed *lin-28* mRNA as well (Seggerson et al. 2002). Subsequently, several other miRNAs (let-7b, and miR-128, -129-2, 326 and -344) were also found associated with polyribosomal fractions in mammalian neurons (Kim et al. 2004; Nelson et al. 2004). However, a recent study concluded that both the lin-4 and let-7 miRNAs in *C. elegans* facilitate inhibition of translation initiation rather than at post-initiation step (Ding and Grosshans 2009). In addition, this report demonstrated that miRNA-mediated mRNA decay often occurs alongside translational inhibition; however, the level of decay varies between miRNA-targeted transcripts (Ding and Grosshans 2009). It is possible that some of the inconsistencies in observations for miRNA-mediated effects could be partially due to the larval stage at which miRNA effects in *C. elegans* were studied, or differences in assays used to measure mRNA decay (i.e., oligo d(T) versus gene-specific oligonucleotides used for qPCR assays) (Ding and Grosshans 2009).

Although several other groups also arrived at the same conclusion as the Ambros lab (i.e., that miRNAs inhibit translation post-initiation (Maroney et al. 2006; Petersen et al. 2006) and do not lead to mRNA degradation (Brennecke et al. 2003; Doench et al. 2003; Zeng et al. 2003)), miRNAs have also demonstrated that they can inhibit translation initiation. Using both the tethering approach and reporters targeted by endogenous let-7 miRNAs they demonstrated that the let-7 miRNA could inhibit translation initiation in HeLa cells (Pillai et al. 2005). In contrast to earlier work, polyribosomal profiling of let-7-targeted mRNAs demonstrated a shift of the targeted mRNA into the upper gradient fractions when the reporter mRNA contained let-7 target sites, consistent with inhibition of translation initiation. Furthermore, cap-independent translation was refractory to miRNA action. This was determined using cap analogs as well as bicistronic constructs containing an IRES or tethered eIF4E or eIF4G initiation factors. Work from the Preiss lab was published soon thereafter, and came to similar conclusions using an artificial miRNA (CXCR4) that targeted a transfected reporter mRNA (Humphreys et al. 2005). In addition, their work also concluded that an mRNA requires both a 5′-cap and 3′ poly(A) tail in order for translation to be efficiently inhibited by the miRNA RISC.

Interestingly, miRNA-mediated translational repression could be derepressed in human cells subject to stress conditions (Bhattacharyya et al. 2006a, 2006b). miR-122 translational repression of endogenous CAT-1 mRNA in Huh7 cells can be reversed upon amino acid deprivation. The translational derepression of CAT-1 mRNA involves a redistribution of the mRNA out of processing bodies and into actively translating polysomes and requires binding of the AU-rich element binding protein HuR to the CAT-1 mRNA 3′UTR. Soon thereafter, the RNA-binding protein Dead end 1 (Dnd1), which binds to U-rich sequences, demonstrated that it too could derepress specific miRNA-targeted mRNAs in both zebrafish and human germ cells (Kedde et al. 2007). Dnd1 binds to U-rich sequences adjacent to miRNA target sites, and interferes with miRNA–RISC access thereby derepressing specific miRNA-targeted mRNAs.

Although the majority of miRNA research has been conducted in vivo, several groups have developed cell-free extracts that recapitulate miRNA repression in vitro. The first in vitro system to be published came from the Novina lab, and was established using a rabbit reticulocyte lysate (Wang et al. 2006, 2008). Using the artificial CXCR4 miRNA and in vitro transcribed reporter mRNAs, they demonstrated that miRNA silencing was cap- and poly(A)-dependent; however, increasing the length of the poly(A) tail seemed to offset the cap-dependence of the system. In addition, using a biotin pulldown approach to capture factors associated with the miRNA-targeted mRNA, they found that targeted mRNAs associated with the 40S ribosomal subunit and eIF2 and eIF3 translation factors, but not with the 60S. Toe printing assays suggested that miRNA-targeted mRNAs exhibited a characteristic 40S subunit toe print, suggesting a miRNA-mediated initiation block at the 60S subunit joining step. A compelling argument in support of a miRNA-mediated block of 60S subunit joining came from the work conducted by the Shiekhattar and Pasquinelli laboratories (Chendrimada et al. 2007). They demonstrated that the 60S ribosomal subunit antiassociation factor eIF6 associates with the human miRISC. They further demonstrated that depletion of eIF6 in either human cells or worms abrogates miRNA-mediated repression (Basu et al. 2001; Sanvito et al. 1999). However, eIF6 has been shown to play a prominent role in 60S subunit biogenesis, complicating interpretation of data regarding a role of eIF6 in the miRNA response.

Several other groups have described additional in vitro systems that recapitulated miRNA-mediated repression. Extracts were made from *D. melanogaster* embryos (Thermann and Hentze 2007), mouse Krebs-2 ascites (Mathonnet et al. 2007), and transfected HEK-293 cells (Wakiyama et al. 2007). Overall, all the three results concluded that miRNAs inhibit translation at the initiation step and that this inhibition is a cap-dependent event. In addition, each group reported unique observations. Experiments with *D. melanogaster* embryonic extracts demonstrated that miR-2 inhibited 40S ribosomal subunit recruitment to the miRNA-targeted mRNA, suggesting a miRNA-mediated block of early events in translation initiation (Thermann and Hentze 2007). In this study, miRNA-targeted mRNAs were associated with "pseudopolysomes" that sedimented faster than 80S ribosomes in a density gradient. Experiments in HEK-293-derived extracts demonstrated that miRNAs induced deadenylation of target mRNAs in vitro (Wakiyama et al. 2007). Work in mouse Krebs-2 extracts demonstrated that addition of recombinant eIF4F antagonized miRNA-mediated translational repression (Mathonnet et al. 2007). This result strongly suggested that early events in translation initiation are targeted by miRNAs (i.e., eIF4F/cap interactions). Kiriakidou et al. presented a compelling hypothesis about how the miRISC might inhibit eIF4F–cap interactions when they demonstrated that the central domains of AGO proteins possess sequence homology to the cap binding region of eIF4E (Kiriakidou et al. 2007). They found that AGO2 binds a cap column and that mutations to two aromatic residues in the central domain of AGO2 blocked its interaction with the cap column. These mutations also inhibited mutant AGO2 translational repression activity when the AGO2 mutant was tethered to the 3′UTR of a reporter mRNA (Kiriakidou et al. 2007). This led to a model whereby AGO2–cap interaction competes with eIF4E for cap binding, thus decreasing the rate of translation initiation. However, a more recent report calls these results into question. Izaurralde and colleagues showed, using *Drosophila* AGO homologues, that the AGO mutant that led to a loss of cap column interaction in the previous study abrogated association of Argonaute with miRNA and with GW182 (a P-body component with affinity for AGO proteins) (Eulalio et al. 2008b). Further, they observed no difference in binding of the mutant *Drosophila* Ago homologue to a cap affinity column. These studies directly contradict one another with regard to cap affinity and miRNA binding capability of mutant Ago. Clearly, additional experiments will be required to determine whether Ago proteins directly interact with the 5′-cap. It is possible that AGO proteins or some other component of the miRISC complex can compete with eIF4E for cap binding. It is also possible that miRNAs exert their effect by preventing proper circularization of the mRNA. It is thought that interaction of PABP with eIF4G, a component of the eIF4F cap binding complex, leads to circularization of mRNAs. This circularization is thought to increase the affinity of eIF4E (the cap binding component of eIF4F) for the cap. Hence, if miRNAs inhibit circularization, this would lead to a loss of affinity of eIF4F for the cap, and hence a decrease in efficiency of translation initiation. This model is consistent with the requirement of both a 5′ cap and a 3′ poly(A) tail to elicit miRNA-mediated repression observed in many systems.

1.3.2   Enhancing Eukaryotic Translation

miRNAs repress gene expression by inhibiting mRNA translation and/or initiating mRNA decay. However recent studies in the Steitz lab suggest that miRNAs may in fact enhance translation, rather than inhibit, under certain cellular conditions (Vasudevan et al. 2007, 2008). Specifically, when mammalian cells are starved of serum, miR369-3 interacts with the 3′UTR of tumor necrosis factor-alpha mRNA and enhances its translation. miRNA-mediated enhancement of translation requires the interaction of FXR1 protein with Ago2. Translational enhancement is not limited to miR369-3, as let-7 and CXCR4 miRNAs also enhanced translation of target mRNAs under serum-starvation conditions. Overall, they present a model whereby miRNAs repress translation in proliferating cells, but enhance translation when cells quiesce. These provocative results introduce a new level of complexity with regard to the mechanism of the miRNA response. It will be interesting to see what cellular events and signaling cascades elicit the switch from miRNA-mediated repression of gene expression to enhancement.

1.3.3   miRNA-Mediated Regulation of mRNA Stability

The original discovery of the lin-4 and let-7 miRNAs in *C. elegans* was accompanied by the demonstration that they inhibit translation without affecting mRNA stability (Lee et al. 1993; Reinhart et al. 2000; Wightman et al. 1993). In contrast to the early reports in *C. elegans*, many groups now observe some level of miRNA-mediated mRNA destabilization. This observation suggests that miRNA-mediated translational repression and mRNA decay act in tandem to facilitate repression of gene expression. This assertion is supported by a report that miRNAs elicit a 95% reduction in reporter expression, as well as a 50% decrease in target mRNA levels (Petersen et al. 2006). In addition, Wu et al. reported that miR-125b expression reduced target protein production by 90%, while mRNA levels were reduced by around 70% (Wu et al. 2006). Hence, translational repression and mRNA destabilization appear to have an additive effect on miRNA-mediated repression of gene expression.

In contrast to siRNA-mediated mRNA endonucleolytic cleavage, miRNA mediated enhancement in the rate of mRNA decay appears to be enacted via more traditional deadenylation-dependent degradation pathways. The first evidence that miRNAs mediate deadenylation of target mRNAs came from work conducted in zebrafish in the Schier laboratory. Studies on clearance of maternal mRNAs following activation of zygotic transcription demonstrated that miR-430 targets a few hundred maternal transcripts and mediates their deadenylation and decay (Giraldez et al. 2006). Studies were also published at the same time demonstrating that miRNAs can induce deadenylation in both *Drosophila melanogaster* S2 cells and as well as in

HeLa and NIH-3T3 cells (Behm-Ansmant et al. 2006; Wu et al. 2006). miRNA mediated deadenylation appears to be mediated by the Caf1–CCR4–Not1 deadenylation complex. Work published by the Izaurralde group (Behm-Ansmant et al. 2006; Eulalio et al. 2008a) demonstrated, in *D. melanogaster* cells, that miRNA-dependent mRNA decay is inhibited by siRNA knockdown of deadenylation factors Not1 and Ccr4 as well as the decapping enzyme Dcp1/2 (Behm-Ansmant et al. 2006). miRNA-mediated deadenylation, in this system, also required the GW182 homolog Gawky, as artificially tethering of Gawky to a reporter 3′UTR stimulated deadenylation in the absence of Ago1 protein, the core Argonaute required for miRNA-mediated deadenylation in *Drosophila.*

Interestingly, miRNA-mediated mRNA degradation and translational repression are suggested to function as independent mechanisms of action. Several groups have demonstrated that mRNAs that are not actively translating can still undergo miRNA-mediated deadenylation and/or decay (Eulalio et al. 2007; Wakiyama et al. 2007). In mammalian cells, miRNA-dependent deadenylation and subsequent complete decay of target mRNA was accelerated by miR-125b expression (Wu et al. 2006). This effect required only a single miR-125b target site and was not affected by inhibition inserting a stable hairpin structure into the reporter mRNA's 5′UTR that prevents translation. miRNA-dependent mRNA deadenylation has also been observed in vitro. Wakiyama et al. reported target mRNA deadenylation in extracts derived from HEK293 cells (Wakiyama et al. 2007). Consistent with in vivo results, deadenylation was not dependent on translation, as mRNAs containing a nonfunctional ApppN cap or IRES-containing mRNAs were subject to deadenylation, despite the fact that these constructs exhibited no translational repression (Wakiyama et al. 2007). mRNA decapping often occurs subsequent to mRNA deadenylation and precedes mRNA decay. Several groups have suggested that miRNAs also function, at least for specific mRNAs, to mediate removal of the 5′-cap structure. Specifically, knocking down decapping factors Dcp1 and/or Dcp2 leads to a stabilization of miRNA-targeted reporter mRNAs (Eulalio et al. 2007; Rehwinkel et al. 2006).

Two recent reports have provided a large-scale picture of miRNA mediated control of both target protein and mRNA levels using mass spectrometric proteomic approaches in parallel with microarray-based analysis of mRNA levels. Selbach et al. introduced five different miRNAs (miR-1, miR-155, miR-16, miR30a, and let7b) into HeLa cells by transfection and also used a locked nucleic acid to knockdown let7b and looked at changes in protein and mRNA levels on a genome-wide scale (Selbach et al. 2008). They report that most targets are repressed at both the mRNA and protein level, with the relative contributions of mRNA destabilization and translation inhibition varying from miRNA to miRNA and from target to target. Interestingly, they found that proteins translated at the endoplasmic reticulum were overrepresented in the class of targets that were repressed mainly at the protein level. Baek et al. used a similar, large-scale approach to look at the effect of transfection of miR-1, miR-124, and miR-181 into HeLa cells as well as the effect of deleting miR-223 from mouse neutrophils (Baek et al. 2008). They found a similar correlation between effects at the mRNA and protein level, reporting that targets exhibiting more than 33% repression at the protein level were also repressed at the mRNA level. Both studies demonstrated the ubiquity of the miRNA response, showing that transfection of single miRNAs generally repressed hundreds of genes at the post-transcriptional level, although few targets were repressed by more than three or fourfold. These results lend credence to the notion that miRNA repression serves to fine-tune gene expression. The Izaurralde lab depleted *D. melanogaster* S2 cells of either AGO1 (the only Argonaute protein involved in the miRNA response in flies), CAF1, or NOT1 and monitored changes in cellular mRNA levels by microarray (Eulalio et al. 2009b). They found that 60% of genes regulated by AGO1 were also regulated by CAF1 and/or NOT1. These results also suggest that mRNA deadenylation plays a significant role in the miRNA response in vivo.

It is clear, then, that miRNA-mediated translational repression and mRNA destabilization act synergistically to inhibit gene expression. mRNA deadenylation removes the binding site for PABP at the mRNA's 3′ end, efficiently disrupting mRNA circularization. As such, deadenylation may be seen as a component of miRNA-mediated translational repression, in addition to its role in initiating decay of the entire mRNA.

1.3.4 The Role of GW182 Proteins and P-Bodies in the miRNA Response

GW182 proteins have recently become a popular topic for studies directed at elucidating mechanistic details of the miRNA response. GW182 is part of a conserved group of proteins characterized by multiple glycine–tryptophan repeat regions that has been found localized to cellular processing bodies (P-bodies, dynamic subcellular structures involved in mRNA storage and decay). GW182 proteins also interact with Argonaute proteins (through GW repeat regions), representing a link between the miRISC and the P-body (Jakymiw et al. 2005; Lian et al. 2009; Liu et al. 2005a, 2005b; Meister et al. 2005; Sen and Blau 2005; Takimoto et al. 2009). It was initially demonstrated in *Drosophila* S2 cells that knockdown of the GW182 homolog *gawky* disrupted miRNA-mediated repression of reporter protein production (Rehwinkel et al. 2005), a result that was reproduced in human cells (Liu et al. 2005a) and in *C. elegans* (Ding and Grosshans 2009). Results from *D. melanogaster* later suggested that miRNA-mediated repression is enacted through GW182 proteins as depletion of Ago1 or Gawky resulted in strikingly similar changes in gene expression by microarray (Behm-Ansmant et al. 2006). Also, tethering of Gawky to the 3′UTR of a reporter gene in the absence of Ago1 resulted in mRNA destabilization as well as a decrease in reporter protein production (Behm-Ansmant et al. 2006), while disruption of Gawky–Ago1 interactions through overexpression of the Ago1 binding domain of Gawky blocked miRNA-mediated silencing of reporters (Eulalio et al. 2008a). A similar result was later reported in human cells (Takimoto et al. 2009). Further evidence for the idea that miRNA effects are mediated through GW182 proteins came from work in human cells. Lian et al. showed that the C-terminal half of all four human Argonaute proteins bind

GW182 and that tethering of this C-terminal half of hAgo2 to the 3′UTR of reporters results in similar levels of repression as tethering full length Ago2 (Lian et al. 2009). Three recent studies have implicated the C-terminus of GW182 proteins as the region responsible for mediating gene silencing. GW182 tethering assays, in concert with deletion analyses, have demonstrated that the C-terminal domain of human GW182 proteins, that cannot bind Argonaute, is sufficient to drive repression of reporter gene expression (Lazzaretti et al. 2009; Zipprich et al. 2009). Importantly, tethering of C-terminal fragments results in repression at both the protein and mRNA levels (Lazzaretti et al. 2009; Zipprich et al. 2009). Genetic analysis in *D. melanogaster* has demonstrated that both the N-terminal Argonaute binding domain and the C-terminal effector domain of Gawky are necessary for miRNA-mediated repression (Eulalio et al. 2009a). Interestingly, a Gawky mutant that fails to localize to P-bodies, but contains the N-terminal Arogonaute binding domain and C-terminal silencing domain, is able to support the miRNA response, while this mutant fails to rescue association of Ago1 to P-bodies (Eulalio et al. 2009a).

Taken together, these results suggest that miRNA mediated translational repression, as well as mRNA destabilization is mediated through GW182 proteins. In effect, the role of the miRNA and Argonaute appears to be to recruit target mRNAs to GW182 proteins, which facilitate translational repression and decay of these transcripts.

1.4 Summary

As data continue to emerge regarding the mechanism of the miRNA response, it has become increasingly apparent that a single concise mechanism cannot account for all examples of miRNA-mediated repression. miRNAs have been reported to repress translation at the level of initiation as well as post-initiation, and to facilitate decay of target mRNAs. It is likely that different cell types, different developmental stages, or different miRNAs may exhibit repression via different mechanisms or combination of mechanisms. One remaining challenge will be to determine what molecular cues determine which mode of repression (or activation) is enacted. Recent reports strongly suggest that miRNA effects are mediated through GW182 proteins and that GW182 effects are not limited to bringing miRNA targeted transcripts to P-bodies. The next step, then, is to dissect the molecular events, downstream of recruitment of GW182 proteins to targeted mRNAs, involved in the various modes of miRNA-mediated control of gene expression.

References

Algire MA, Maag D, Lorsch JR (2005) Pi release from eIF2, not GTP hydrolysis, is the step controlled by start-site selection during eukaryotic translation initiation. Mol Cell 20:251–262

Baek D, Villen J, Shin C, Camargo FD, Gygi SP, Bartel DP (2008) The impact of microRNAs on protein output. Nature 455:64–71

Bagga S, Bracht J, Hunter S, Massirer K, Holtz J, Eachus R, Pasquinelli AE (2005) Regulation by let-7 and lin-4 miRNAs results in target mRNA degradation. Cell 122:553–563

Bartel DP (2004) MicroRNAs: genomics, biogenesis, mechanism, and function. Cell 116:281–297

Basu U, Si K, Warner JR, Maitra U (2001) The Saccharomyces cerevisiae TIF6 gene encoding translation initiation factor 6 is required for 60S ribosomal subunit biogenesis. Mol Cell Biol 21:1453–1462

Behm-Ansmant I, Rehwinkel J, Doerks T, Stark A, Bork P, Izaurralde E (2006) mRNA degradation by miRNAs and GW182 requires both CCR4:NOT deadenylase and DCP1:DCP2 decapping complexes. Genes Dev 20:1885–1898

Berezikov E, Chung WJ, Willis J, Cuppen E, Lai EC (2007) Mammalian mirtron genes. Mol Cell 28:328–336

Bhattacharyya SN, Habermacher R, Martine U, Closs EI, Filipowicz W (2006a) Relief of microRNA-mediated translational repression in human cells subjected to stress. Cell 125:1111–1124

Bhattacharyya SN, Habermacher R, Martine U, Closs EI, Filipowicz W (2006b) Stress-induced reversal of microRNA repression and mRNA P-body localization in human cells. Cold Spring Harb Symp Quant Biol 71:513–521

Bi X, Ren J, Goss DJ (2000) Wheat germ translation initiation factor eIF4B affects eIF4A and eIFiso4F helicase activity by increasing the ATP binding affinity of eIF4A. Biochemistry 39:5758–5765

Brennecke J, Hipfner DR, Stark A, Russell RB, Cohen SM (2003) Bantam encodes a developmentally regulated microRNA that controls cell proliferation and regulates the proapoptotic gene hid in Drosophila. Cell 113:25–36

Calin GA, Croce CM (2006a) MicroRNA-cancer connection: the beginning of a new tale. Cancer Res 66:7390–7394

Calin GA, Croce CM (2006b) MicroRNA signatures in human cancers. Nat Rev Cancer 6:857–866

Ceci M, Gaviraghi C, Gorrini C, Sala LA, Offenhauser N, Marchisio PC, Biffo S (2003) Release of eIF6 (p27BBP) from the 60S subunit allows 80S ribosome assembly. Nature 426:579–584

Chang TC, Mendell JT (2007) microRNAs in vertebrate physiology and human disease. Annu Rev Genom Hum Genet 8:215–239

Chen CZ, Lodish HF (2005) MicroRNAs as regulators of mammalian hematopoiesis. Semin Immunol 17:155–165

Chendrimada TP, Finn KJ, Ji X, Baillat D, Gregory RI, Liebhaber SA, Pasquinelli AE, Shiekhattar R (2007) MicroRNA silencing through RISC recruitment of eIF6. Nature 447:823–828

Croce CM, Calin GA (2005) miRNAs, cancer, and stem cell division. Cell 122:6–7

Cullen BR (2009) Viral and cellular messenger RNA targets of viral microRNAs. Nature 457:421–425

Cummins JM, Velculescu VE (2006) Implications of micro-RNA profiling for cancer diagnosis. Oncogene 25:6220–6227

Dalmay T, Edwards DR (2006) MicroRNAs and the hallmarks of cancer. Oncogene 25:6170–6175

Ding XC, Grosshans H (2009) Repression of C. elegans microRNA targets at the initiation level of translation requires GW182 proteins. EMBO J 28:213–222

Doench JG, Sharp PA (2004) Specificity of microRNA target selection in translational repression. Genes Dev 18:504–511

Doench JG, Petersen CP, Sharp PA (2003) siRNAs can function as miRNAs. Genes Dev 17:438–442

Edery I, Humbelin M, Darveau A, Lee KA, Milburn S, Hershey JW, Trachsel H, Sonenberg N (1983) Involvement of eukaryotic initiation factor 4A in the cap recognition process. J Biol Chem 258:11398–11403

Eulalio A, Rehwinkel J, Stricker M, Huntzinger E, Yang SF, Doerks T, Dorner S, Bork P, Boutros M, Izaurralde E (2007) Target-specific requirements for enhancers of decapping in miRNA-mediated gene silencing. Genes Dev 21:2558–2570

Eulalio A, Huntzinger E, Izaurralde E (2008a) GW182 interaction with Argonaute is essential for miRNA-mediated translational repression and mRNA decay. Nat Struct Mol Biol 15 :346–353

Eulalio A, Huntzinger E, Izaurralde E (2008b) GW182 interaction with Argonaute is essential for miRNA-mediated translational repression and mRNA decay. Nat Struct Mol Biol

Eulalio A, Helms S, Fritzsch C, Fauser M, Izaurralde E (2009a) A C-terminal silencing domain in GW182 is essential for miRNA function. RNA

Eulalio A, Huntzinger E, Nishihara T, Rehwinkel J, Fauser M, Izaurralde E (2009b) Deadenylation is a widespread effect of miRNA regulation. RNA 15:21–32

Forman JJ, Legesse-Miller A, Coller HA (2008) A search for conserved sequences in coding regions reveals that the let-7 microRNA targets Dicer within its coding sequence. Proc Natl Acad Sci U S A 105:14879–14884

Gallie DR (1991) The cap and poly(A) tail function synergistically to regulate mRNA translational efficiency. Genes Dev 5:2108–2116

Garzon R, Fabbri M, Cimmino A, Calin GA, Croce CM (2006) MicroRNA expression and function in cancer. Trends Mol Med 12:580–587

Giannakakis A, Coukos G, Hatzigeorgiou A, Sandaltzopoulos R, Zhang L (2007) miRNA genetic alterations in human cancers. Expert Opin Biol Ther 7:1375–1386

Gilbert WV, Zhou K, Butler TK, Doudna JA (2007) Cap-independent translation is required for starvation-induced differentiation in yeast. Science 317:1224–1227

Giraldez AJ, Mishima Y, Rihel J, Grocock RJ, Van Dongen S, Inoue K, Enright AJ, Schier AF (2006) Zebrafish MiR-430 promotes deadenylation and clearance of maternal mRNAs. Science 312:75–79

Grifo JA, Tahara SM, Morgan MA, Shatkin AJ, Merrick WC (1983) New initiation factor activity required for globin mRNA translation. J Biol Chem 258:5804–5810

Gu S, Jin L, Zhang F, Sarnow P, Kay MA (2009) Biological basis for restriction of microRNA targets to the 3' untranslated region in mammalian mRNAs. Nat Struct Mol Biol 16:144–150

Hammond SM (2006) MicroRNAs as oncogenes. Curr Opin Genet Dev 16:4–9

Han J, Lee Y, Yeom KH, Kim YK, Jin H, Kim VN (2004) The Drosha-DGCR8 complex in primary microRNA processing. Genes Dev 18:3016–3027

Han J, Lee Y, Yeom KH, Nam JW, Heo I, Rhee JK, Sohn SY, Cho Y, Zhang BT, Kim VN (2006) Molecular basis for the recognition of primary microRNAs by the Drosha-DGCR8 complex. Cell 125:887–901

He L, He X, Lowe SW, Hannon GJ (2007) microRNAs join the p53 network–another piece in the tumour-suppression puzzle. Nat Rev Cancer 7:819–822

Hellen CU, Wimmer E (1995) Translation of encephalomyocarditis virus RNA by internal ribosomal entry. Curr Top Microbiol Immunol 203:31–63

Henis-Korenblit S, Strumpf NL, Goldstaub D, Kimchi A (2000) A novel form of DAP5 protein accumulates in apoptotic cells as a result of caspase cleavage and internal ribosome entry site-mediated translation. Mol Cell Biol 20:496–506

Holcik M, Lefebvre C, Yeh C, Chow T, Korneluk RG (1999) A new internal-ribosome-entry-site motif potentiates XIAP-mediated cytoprotection. Nat Cell Biol 1:190–192

Humphreys DT, Westman BJ, Martin DI, Preiss T (2005) MicroRNAs control translation initiation by inhibiting eukaryotic initiation factor 4E/cap and poly(A) tail function. Proc Natl Acad Sci U S A 102:16961–16966

Imataka H, Olsen HS, Sonenberg N (1997) A new translational regulator with homology to eukaryotic translation initiation factor 4G. EMBO J 16:817–825

Jakymiw A, Lian S, Eystathioy T, Li S, Satoh M, Hamel JC, Fritzler MJ, Chan EK (2005) Disruption of GW bodies impairs mammalian RNA interference. Nat Cell Biol 7:1267–1274

Jan E, Sarnow P (2002) Factorless ribosome assembly on the internal ribosome entry site of cricket paralysis virus. J Mol Biol 324:889–902

Jang SK, Krausslich HG, Nicklin MJ, Duke GM, Palmenberg AC, Wimmer E (1988) A segment of the 5' nontranslated region of encephalomyocarditis virus RNA directs internal entry of ribosomes during in vitro translation. J Virol 62:2636–2643

Kataoka N, Fujita M, Ohno M (2009) Functional association of the Microprocessor complex with spliceosome. Mol Cell Biol

Kedde M, Strasser MJ, Boldajipour B, Oude Vrielink JA, Slanchev K, le Sage C, Nagel R, Voorhoeve PM, van Duijse J, Orom UA, Lund AH, Perrakis A, Raz E, Agami R (2007) RNA-binding protein Dnd1 inhibits microRNA access to target mRNA. Cell 131:1273–1286

Kim YK, Kim VN (2007) Processing of intronic microRNAs. EMBO J 26:775–783

Kim J, Krichevsky A, Grad Y, Hayes GD, Kosik KS, Church GM, Ruvkun G (2004) Identification of many microRNAs that copurify with polyribosomes in mammalian neurons. Proc Natl Acad Sci U S A 101:360–365

Kiriakidou M, Tan GS, Lamprinaki S, De Planell-Saguer M, Nelson PT, Mourelatos Z (2007) An mRNA m7G cap binding-like motif within human Ago2 represses translation. Cell 129: 1141–1151

Kolupaeva VG, Pestova TV, Hellen CU, Shatsky IN (1998) Translation eukaryotic initiation factor 4G recognizes a specific structural element within the internal ribosome entry site of encephalomyocarditis virus RNA. J Biol Chem 273:18599–18604

Kozak M (1978) How do eucaryotic ribosomes select initiation regions in messenger RNA? Cell 15:1109–1123

Kozak M, Shatkin AJ (1979) Characterization of translational initiation regions from eukaryotic messenger RNAs. Methods Enzymol 60:360–375

Lazzaretti D, Tournier I, Izaurralde E (2009) The C-terminal domains of human TNRC6A, TNRC6B, and TNRC6C silence bound transcripts independently of Argonaute proteins. RNA

Lee RC, Ambros V (2001) An extensive class of small RNAs in Caenorhabditis elegans. Science 294:862–864

Lee RC, Feinbaum RL, Ambros V (1993) The C. elegans heterochronic gene lin-4 encodes small RNAs with antisense complementarity to lin-14. Cell 75:843–854

Lian SL, Li S, Abadal GX, Pauley BA, Fritzler MJ, Chan EK (2009) The C-terminal half of human Ago2 binds to multiple GW-rich regions of GW182 and requires GW182 to mediate silencing. RNA 15:804–813

Ling J, Morley SJ, Pain VM, Marzluff WF, Gallie DR (2002) The histone 3'-terminal stem-loop-binding protein enhances translation through a functional and physical interaction with eukaryotic initiation factor 4G (eIF4G) and eIF3. Mol Cell Biol 22:7853–7867

Liu J, Rivas FV, Wohlschlegel J, Yates JR 3rd, Parker R, Hannon GJ (2005a) A role for the P-body component GW182 in microRNA function. Nat Cell Biol 7:1261–1266

Liu J, Valencia-Sanchez MA, Hannon GJ, Parker R (2005b) MicroRNA-dependent localization of targeted mRNAs to mammalian P-bodies. Nat Cell Biol 7:719–723

Lytle JR, Yario TA, Steitz JA (2007) Target mRNAs are repressed as efficiently by microRNA-binding sites in the 5' UTR as in the 3' UTR. Proc Natl Acad Sci U S A 104:9667–9672

Maag D, Fekete CA, Gryczynski Z, Lorsch JR (2005) A conformational change in the eukaryotic translation preinitiation complex and release of eIF1 signal recognition of the start codon. Mol Cell 17:265–275

Maroney PA, Yu Y, Fisher J, Nilsen TW (2006) Evidence that microRNAs are associated with translating messenger RNAs in human cells. Nat Struct Mol Biol 13:1102–1107

Mathonnet G, Fabian MR, Svitkin YV, Parsyan A, Huck L, Murata T, Biffo S, Merrick WC, Darzynkiewicz E, Pillai RS, Filipowicz W, Duchaine TF, Sonenberg N (2007) MicroRNA inhibition of translation initiation in vitro by targeting the cap-binding complex eIF4F. Science 317:1764–1767

Mattes J, Collison A, Foster PS (2008) Emerging role of microRNAs in disease pathogenesis and strategies for therapeutic modulation. Curr Opin Mol Ther 10:150–157

Meister G, Landthaler M, Peters L, Chen PY, Urlaub H, Luhrmann R, Tuschl T (2005) Identification of novel argonaute-associated proteins. Curr Biol 15:2149–2155

Mitchell SF, Lorsch JR (2008) Should I stay or should I go? Eukaryotic translation initiation factors 1 and 1A control start codon recognition. J Biol Chem 283:27345–27349

Nelson PT, Hatzigeorgiou AG, Mourelatos Z (2004) miRNP:mRNA association in polyribosomes in a human neuronal cell line. RNA 10:387–394

Okamura K, Hagen JW, Duan H, Tyler DM, Lai EC (2007) The mirtron pathway generates microRNA-class regulatory RNAs in Drosophila. Cell 130:89–100

Olsen PH, Ambros V (1999) The lin-4 regulatory RNA controls developmental timing in Caenorhabditis elegans by blocking LIN-14 protein synthesis after the initiation of translation. Dev Biol 216:671–680

Orom UA, Nielsen FC, Lund AH (2008) MicroRNA-10a binds the 5'UTR of ribosomal protein mRNAs and enhances their translation. Mol Cell 30:460–471

Pasquinelli AE, Reinhart BJ, Slack F, Martindale MQ, Kuroda MI, Maller B, Hayward DC, Ball EE, Degnan B, Muller P, Spring J, Srinivasan A, Fishman M, Finnerty J, Corbo J, Levine M, Leahy P, Davidson E, Ruvkun G (2000) Conservation of the sequence and temporal expression of let-7 heterochronic regulatory RNA. Nature 408:86–89

Pelletier J, Sonenberg N (1988) Internal initiation of translation of eukaryotic mRNA directed by a sequence derived from poliovirus RNA. Nature 334:320–325

Pestova TV, Kolupaeva VG (2002) The roles of individual eukaryotic translation initiation factors in ribosomal scanning and initiation codon selection. Genes Dev 16:2906–2922

Pestova TV, Lomakin IB, Lee JH, Choi SK, Dever TE, Hellen CU (2000) The joining of ribosomal subunits in eukaryotes requires eIF5B. Nature 403:332–335

Petersen CP, Bordeleau ME, Pelletier J, Sharp PA (2006) Short RNAs repress translation after initiation in mammalian cells. Mol Cell 21:533–542

Pillai RS, Artus CG, Filipowicz W (2004) Tethering of human Ago proteins to mRNA mimics the miRNA-mediated repression of protein synthesis. RNA 10:1518–1525

Pillai RS, Bhattacharyya SN, Artus CG, Zoller T, Cougot N, Basyuk E, Bertrand E, Filipowicz W (2005) Inhibition of translational initiation by Let-7 MicroRNA in human cells. Science 309:1573–1576

Pisarev AV, Shirokikh NE, Hellen CU (2005) Translation initiation by factor-independent binding of eukaryotic ribosomes to internal ribosomal entry sites. C R Biol 328:589–605

Poy MN, Eliasson L, Krutzfeldt J, Kuwajima S, Ma X, Macdonald PE, Pfeffer S, Tuschl T, Rajewsky N, Rorsman P, Stoffel M (2004) A pancreatic islet-specific microRNA regulates insulin secretion. Nature 432:226–230

Rehwinkel J, Behm-Ansmant I, Gatfield D, Izaurralde E (2005) A crucial role for GW182 and the DCP1:DCP2 decapping complex in miRNA-mediated gene silencing. RNA 11:1640–1647

Rehwinkel J, Natalin P, Stark A, Brennecke J, Cohen SM, Izaurralde E (2006) Genome-wide analysis of mRNAs regulated by Drosha and Argonaute proteins in Drosophila melanogaster. Mol Cell Biol 26:2965–2975

Reinhart BJ, Slack FJ, Basson M, Pasquinelli AE, Bettinger JC, Rougvie AE, Horvitz HR, Ruvkun G (2000) The 21-nucleotide let-7 RNA regulates developmental timing in Caenorhabditis elegans. Nature 403:901–906

Rigoutsos I (2009) New tricks for animal microRNAS: targeting of amino acid coding regions at conserved and nonconserved sites. Cancer Res 69:3245–3248

Rogers GW Jr, Richter NJ, Merrick WC (1999) Biochemical and kinetic characterization of the RNA helicase activity of eukaryotic initiation factor 4A. J Biol Chem 274:12236–12244

Ruby JG, Jan CH, Bartel DP (2007) Intronic microRNA precursors that bypass Drosha processing. Nature 448:83–86

Sachs A (2000) Physical and functional interactions between the mRNA cap structure and the poly(A) tail. In Sonenberg N, Hershey JWB, Mathews MB (eds) Translational control of gene expression. Cold Spring Harbor Laboratory, Cold Spring Harbor, NY, pp 447–466

Sachs AB, Varani G (2000) Eukaryotic translation initiation: there are (at least) two sides to every story. Nat Struct Biol 7:356–361

Sanvito F, Piatti S, Villa A, Bossi M, Lucchini G, Marchisio PC, Biffo S (1999) The beta4 integrin interactor p27(BBP/eIF6) is an essential nuclear matrix protein involved in 60S ribosomal subunit assembly. J Cell Biol 144:823–837

Seggerson K, Tang L, Moss EG (2002) Two genetic circuits repress the Caenorhabditis elegans heterochronic gene lin-28 after translation initiation. Dev Biol 243:215–225

Selbach M, Schwanhausser B, Thierfelder N, Fang Z, Khanin R, Rajewsky N (2008) Widespread changes in protein synthesis induced by microRNAs. Nature 455:58–63

Sen GL, Blau HM (2005) Argonaute 2/RISC resides in sites of mammalian mRNA decay known as cytoplasmic bodies. Nat Cell Biol 7:633–636

Sonenberg N, Rupprecht KM, Hecht SM, Shatkin AJ (1979) Eukaryotic mRNA cap binding protein: purification by affinity chromatography on sepharose-coupled m7GDP. Proc Natl Acad Sci U S A 76:4345–4349

Spahn CM, Jan E, Mulder A, Grassucci RA, Sarnow P, Frank J (2004) Cryo-EM visualization of a viral internal ribosome entry site bound to human ribosomes: the IRES functions as an RNA-based translation factor. Cell 118:465–475

Stark A, Brennecke J, Bushati N, Russell RB, Cohen SM (2005) Animal MicroRNAs confer robustness to gene expression and have a significant impact on 3'UTR evolution. Cell 123:1133–1146

Stefani G (2007) Roles of microRNAs and their targets in cancer. Expert Opin Biol Ther 7:1833–1840

Stoneley M, Paulin FE, Le Quesne JP, Chappell SA, Willis AE (1998) C-Myc 5' untranslated region contains an internal ribosome entry segment. Oncogene 16:423–428

Takimoto K, Wakiyama M, Yokoyama S (2009) Mammalian GW182 contains multiple Argonaute-binding sites and functions in microRNA-mediated translational repression. RNA

Thermann R, Hentze MW (2007) Drosophila miR2 induces pseudo-polysomes and inhibits translation initiation. Nature 447:875–878

Vasudevan S, Tong Y, Steitz JA (2007) Switching from repression to activation: microRNAs can up-regulate translation. Science 318:1931–1934

Vasudevan S, Tong Y, Steitz JA (2008) Cell-cycle control of microRNA-mediated translation regulation. Cell Cyc 7:1545–1549

Wakiyama M, Takimoto K, Ohara O, Yokoyama S (2007) Let-7 microRNA-mediated mRNA deadenylation and translational repression in a mammalian cell-free system. Genes Dev 21:1857–1862

Wang B, Love TM, Call ME, Doench JG, Novina CD (2006) Recapitulation of short RNA-directed translational gene silencing in vitro. Mol Cell 22:553–560

Wang B, Yanez A, Novina CD (2008) MicroRNA-repressed mRNAs contain 40S but not 60S components. Proc Natl Acad Sci U S A 105:5343–5348

Welch C, Chen Y, Stallings RL (2007) MicroRNA-34a functions as a potential tumor suppressor by inducing apoptosis in neuroblastoma cells. Oncogene 26:5017–5022

Wickens M, Goodwin EB, Kimble J, Strickland S, Hentze MW (2000) Translational control of developmental decisions. In Sonenberg N, Hershey JWB, Mathews MB (eds) Translational control of gene expression. Cold Spring Harbor Laboratory, Cold Spring Harbor, NY, pp 295–370

Wightman B, Burglin TR, Gatto J, Arasu P, Ruvkun G (1991) Negative regulatory sequences in the lin-14 3'-untranslated region are necessary to generate a temporal switch during Caenorhabditis elegans development. Genes Dev 5:1813–1824

Wightman B, Ha I, Ruvkun G (1993) Posttranscriptional regulation of the heterochronic gene lin-14 by lin-4 mediates temporal pattern formation in C. elegans. Cell 75:855–862

Wilson JE, Powell MJ, Hoover SE, Sarnow P (2000) Naturally occurring dicistronic cricket paralysis virus RNA is regulated by two internal ribosome entry sites. Mol Cell Biol 20:4990–4999

Wu L, Fan J, Belasco JG (2006) MicroRNAs direct rapid deadenylation of mRNA. Proc Natl Acad Sci U S A 103:4034–4039

Yi R, Qin Y, Macara IG, Cullen BR (2003) Exportin-5 mediates the nuclear export of pre-microRNAs and short hairpin RNAs. Genes Dev 17:3011–3016

Zeng Y, Cullen BR (2003) Sequence requirements for micro RNA processing and function in human cells. RNA 9:112–123

Zeng Y, Cullen BR (2005) Efficient processing of primary microRNA hairpins by Drosha requires flanking nonstructured RNA sequences. J Biol Chem 280:27595–27603

Zeng Y, Yi R, Cullen BR (2003) MicroRNAs and small interfering RNAs can inhibit mRNA expression by similar mechanisms. Proc Natl Acad Sci U S A 100:9779–9784

Zipprich JT, Bhattacharyya S, Mathys H, Filipowicz W (2009) Importance of the C-terminal domain of the human GW182 protein TNRC6C for translational repression. RNA, Epub March 20, 2009

Chapter 2
Translational Control of Endogenous MicroRNA Target Genes in *C. elegans*

Benjamin A. Hurschler, Xavier C. Ding, and Helge Großhans

Contents

Abstract *lin-4* and *let-7* are the founding members of the large microRNA (miRNA) family of regulatory RNAs and were originally identified as components of a *C. elegans* developmental pathway that controls temporal cell fates. Consistent with their pioneering role, *lin-4* and *let-7* were studied widely as "model miR-NAs" in efforts to reveal the mode of action of miRNAs. Early work on *lin-4* thus established a paradigm that miRNAs inhibit translation of their target mRNAs at a step downstream from initiation, without affecting mRNA stability. Although some studies on mammalian miRNAs in cell culture reached similar conclusions, most of those studies indicated that miRNAs repressed translation initiation and frequently also promoted target mRNA degradation. We will discuss here what is known about

B.A. Hurschler, X.C. Ding, and H. Großhans (✉)
Friedrich Miescher Institute for Biomedical Research (FMI), Maulbeerstrasse 66,
WRO-1066.1.38, CH-4002 Basel, Switzerland
e-mail: helge.großhans@fmi.ch

R.E. Rhoads (ed.), *miRNA Regulation of the Translational Machinery*,
Progress in Molecular and Subcellular Biology 50,
DOI 10.1007/978-3-642-03103-8_2, © Springer-Verlag Berlin Heidelberg 2010

modes of miRNA target gene repression in *C. elegans*, highlighting recent work that demonstrates that both mRNA degradation and repression of translation initiation are mechanisms employed in vivo by *let-7* and, unexpectedly, *lin-4* to silence their endogenous targets. We will also discuss the roles of the GW182 homologous AIN-1 and AIN-2 proteins in this process.

2.1 Introduction

lin-4 and *let-7* are the founding members of the large microRNA (miRNA) family of small noncoding RNAs and were originally identified as components of the heterochronic developmental pathway in the small roundworm *Caenorhabditis elegans* (Chalfie et al. 1981; Horvitz and Sulston 1980). *C. elegans* genetics has also been instrumental in the identification of the first miRNA target genes (Moss et al. 1997; Slack et al. 2000; Wightman et al. 1993) and the cellular machinery involved in miRNA mediated gene silencing, e.g., the RNase III enzyme DCR-1 (Dicer) (Grishok et al. 2001; Ketting et al. 2001; Knight and Bass 2001), the Argonaute-like proteins ALG-1, ALG-2 (Grishok et al. 2001), and the microprocessor complex (Denli et al. 2004). Findings in *C. elegans* have thus had a remarkable track record of guiding our understanding of miRNA biology. Indeed, the earliest work on the mechanism of action used by miRNAs to silence their target mRNAs was also performed in *C. elegans* (Olsen and Ambros 1999; Seggerson et al. 2002). It established a paradigm that miRNAs inhibited protein translation at a step downstream of initiation, without significantly affecting target mRNA stability. Surprisingly then, work in human and Drosophila cells has challenged this model of miRNA activity, by providing evidence for miRNA-mediated transcript degradation as well as repression of translation initiation. In this chapter, we discuss what is known about modes of miRNA target gene repression in *C. elegans* and how this relates to findings from other model systems. We particularly focus on recent work that demonstrates that *let-7* and *lin-4* employ both mRNA degradation and, unexpectedly, repression of translation initiation to silence their endogenous targets in vivo. We also discuss the roles of the GW182 homologous AIN-1 and AIN-2 proteins in these processes.

2.2 *lin-4* and *let-7* miRNAs in *C. elegans* Development

Postembryonic development of *C. elegans* proceeds through four larval stages, L1 through L4, each separated by a molt, until the sexually mature adult stage is reached. In a newly hatched larva, 51 blast cells divide and differentiate in a stereotypic manner during the four larval stages, giving rise to a fixed number of cells with determined fates. Proper temporal execution of cell fates is controlled by a set of heterochronic genes. Mutations in these genes can cause either a precocious phenotype, in which developmental events are skipped, or a retarded phenotype, in which developmental events are repeated. For instance, loss-of-function in *lin-4* (*lineage variant-4*) causes reiteration of first larval stage cell fates during the second larval stage in various tissues, whereas mutations in *lin-14* cause a skipping of L1 cell fates (Moss 2007). Surprisingly, *lin-4* was found to code not for a protein, but for a small RNA, capable of triggering L2 fates by diminishing the protein levels of LIN-14 (Lee et al. 1993; Wightman et al. 1993) and LIN-28 (Moss et al. 1997) (Fig. 2.1). *lin-4* achieved repression of the *lin-14* and *lin-28* mRNAs by binding to complementary sequences in their 3′ untranslated regions (3′ UTRs) (Lee et al. 1993; Moss et al. 1997; Wightman et al. 1993).

Seven years later it was discovered that another heterochronic gene, *let-7* (*lethal-7*), also encoded for a small regulatory RNA that regulated temporal cell fates, in this case by promoting transition from L4 to adult cell fates through repression of *lin-41* (Reinhart et al. 2000; Slack et al. 2000). Due to their temporally regulated levels and their function as temporal switches for cell fates in *C. elegans*, *lin-4*

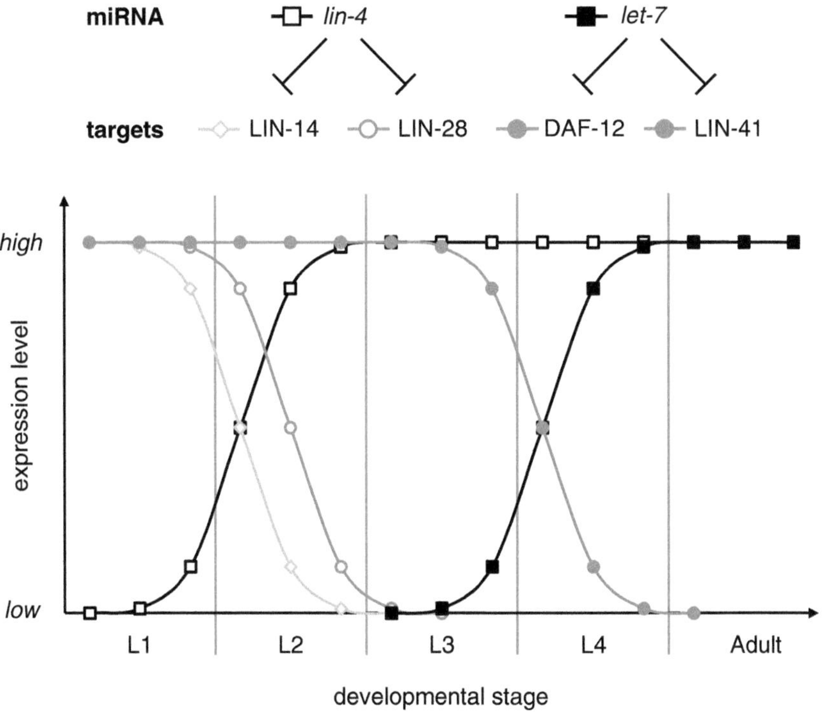

Fig. 2.1 Temporal expression of heterochronic genes in *C. elegans*. The postembryonic development in *C. elegans* proceeds through four larval stages (L1 to L4), each separated by a molt (*indicated by the vertical lines*), followed by the adult stage. *lin-4* starts to accumulate during L1 and represses *lin-14* starting mid-L1 and *lin-28* starting late L1/early L2, thereby promoting progression to developmental programs of L2 and L3, respectively. *let-7* starts to accumulate during L3 and represses *lin-41* and *daf-12* starting late L3/early L4, thereby promoting progression to adult cell-fates

and *let-7* were termed small temporal RNAs. Subsequently, homologues of *let-7* were identified in a variety of bilaterian species, including flies, zebrafish, and humans (Pasquinelli et al. 2000). It was this discovery that provided the starting point for the subsequent isolation of hundreds of miRNAs in various animals, including humans (reviewed in Großhans and Slack 2002).

2.3 Polysome Profiling as an Assay to Assess the Translational State of mRNAs

The discovery that *lin-4* was partially complementary to sequences in the 3′ UTR of the *lin-14* mRNA and that these 3′ UTR sequences were required for regulation (Lee et al. 1993; Wightman et al. 1993) suggested that miRNAs regulate their targets through an antisense mechanism, possibly inducing mRNA degradation or translational repression. Although transcript degradation can be readily assessed by diverse techniques such as northern blotting, quantitative reverse transcription PCR (qRT-PCR), or microarrays, the appraisal of the translational state of a transcript is less straight-forward. Based on the observation that actively trans-lated mRNAs are bound by many ribosomes, isolation of polyribosomes ("poly-somes") can be used to copurify translated mRNAs. The prevalent method for the isolation of polysomes dates back to the early days of studies on protein translation (Wettstein et al. 1963). In its basic implementation, the transcripts in a cleared cell lysate (i.e., the postmitochondrial supernatant) are separated by ultracentrifugation through a sucrose density gradient. While the gradient is unloaded at a constant flow-rate, the UV-absorbance is recorded and fractions are collected. mRNAs that are associated with multiple ribosomes migrate to the denser fractions of the gradi-ent, which can be observed on the UV-recording as a pattern of density peaks cor-responding to multiples of 80S (Fig. 2.2). The 80S peak thus delimits the polysomal and the (sub)monosomal fractions. RNA can then be extracted from polysomal and (sub)monosomal fractions and analyzed by any quantitative assay, e.g., qRT-PCR and northern blotting. Different mRNAs will vary in their distributions across these fractions, reflecting for instance the fact that the number of ribosomes that can be loaded onto short transcripts is limited, but each tran-script exhibits a characteristic, invariant distribution under constant experimental conditions. By contrast, if experimental conditions change to cause, for instance, activation of translation initiation, an increased accumulation in polysomal frac-tions results for the affected transcripts, whereas inhibition of translation initation will cause a shift to (sub)monosomal fractions. To "freeze" polysomes for the duration of the experiment, cells are typically treated with cycloheximide, which blocks elongation of the nascent polypeptide chain. A frequently used control is the application of puromycin, which induces premature termination of translation, and thus specifically disassembles actively translating polysomes, resulting in a shift of the associated mRNA.

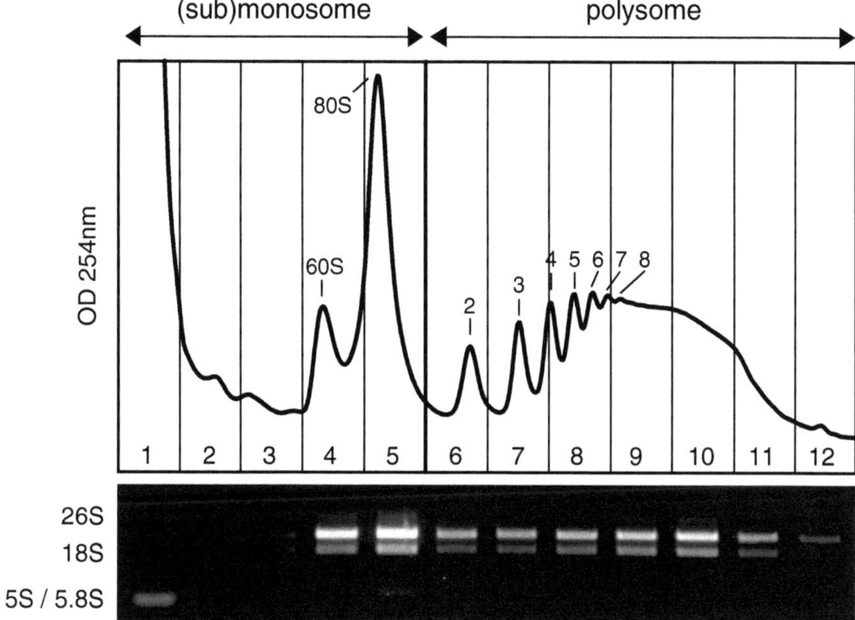

From Ding and Großhans, 2009

Fig. 2.2 A typical polysome profile. UV-recording at 254 nm of total worm lysate separated on a 15–60% (w/v) sucrose gradient. The major UV density peaks represent the 60S sub-monosomal, 80S monosomal, and a series of polysomal peaks (*from left to right*; the number of ribosomes is indicated). Total RNA of each fraction was isolated and separated on an agarose gel to visualize ribosomal RNAs. Adapted from (Ding and Großhans 2009)

2.4   MicroRNA-Mediated Gene Regulation in *C. elegans*: The Early View

Early work performed in *C. elegans* on *lin-4* and *let-7* established an antisense mechanism of interaction between miRNAs and target mRNAs. Gain-of-function mutations of *lin-14* yielded retarded phenotypes resembling those seen with *lin-4* loss-of-function and were caused by deletions in the 3′ UTR of *lin-14*. In both these mutant animals, LIN-14 protein persisted at a developmental stage, in which the protein was no longer detectable in wild-type animals (Olsen and Ambros 1999; Wightman et al. 1993). Reporter gene experiments then confirmed that the 3′ UTR of *lin-14* was sufficient for gene repression by *lin-4*, with mutations in the *lin-4* complementary regions compromising reporter gene regulation (Wightman et al. 1993). The mechanism of regulation however remained elusive. The massive fold decrease in LIN-14 protein between L1 and L2 was not adequately reflected by a decline in the transcript level, and the polyadenylation state of *lin-14* was not affected. Furthermore, *lin-14* was found to cosediment with actively transcribing polysomes in sucrose density gradients both before and after the onset of *lin-4* expression (Olsen and Ambros 1999). Since *lin-14* did not exhibit a shift to the submonosomal fraction, a hallmark of repressed translation initiation, it was concluded that *lin-4* regulated *lin-14* downstream of translation initiation or even posttranslationally. Moreover, a subset of *lin-4* was found to comigrate with polysomes, a finding that was consistent with, although not necessarily diagnostic of, regulation after the initiation step (cf. Sect. 2.5.1).

Similar results were also obtained for another *lin-4* target, *lin-28* (Seggerson et al. 2002), which fostered the paradigm of miRNAs inhibiting translation at a step downstream of initiation, without substantially affecting mRNA stability.

However, more recent work provides evidence for miRNA-dependent target mRNA decay in *C. elegans* (Bagga et al. 2005), an observation that is consistent with a large body of work from other systems (Behm-Ansmant et al. 2006; Eulalio et al. 2007b; Giraldez et al. 2006; Wu and Belasco 2005). Northern blots of endogenous *C. elegans* mRNAs showed a more than fivefold decrease in the *lin-4* targets *lin-14* and *lin-28*, which was more than previously appreciated and *let-7* was similarly found to mediate degradation of its target *lin-41* (Bagga et al. 2005). To explain the discrepancy, it was speculated (Bagga et al. 2005) that previous studies with *C. elegans* (Olsen and Ambros 1999; Seggerson et al. 2002; Wightman et al. 1993), which were based on RNase protection experiments, were distorted by the detection of stable degradation products, but no such degradation intermediates have been demonstrated. We have recently shown transcript degradation for additional *C. elegans* miRNA targets and demonstrated that *C. elegans* miRNAs also block translation initiation (Ding and Großhans 2009) (see Sects. 2.6 and 2.7). Although some evidence suggests that degradation and translational repression are two distinct modes of miRNA target gene repression, it is still possible that degradation may indeed be a consequence of translational repression.

2.5 MicroRNA Mediated Gene Regulation in Other Model Organisms

Many in vivo and in vitro studies have been performed to elucidate the mechanism(s) of miRNA-mediated gene repression in different experimental systems. The resulting plethora of proposed mechanisms of action has sparked a lively debate that characterizes the field. We will shortly review some of the major findings (and conflicts among them), mostly obtained using cell-based assays and reporter genes, before we will discuss recent results on the mechanisms *C. elegans* miRNAs utilize to silence endogenous target genes in vivo, in a whole organism.

2.5.1 Evidence for Translational Repression After Initiation

Several cell-based (ex vivo) studies report translational repression after initiation, although they differ in their conclusion as to how this regulation takes place. In 293T cells, transfection of an artificial miRNA repressed its target reporter mRNA, which remained associated with actively translating polysomes. (Petersen et al. 2006). Repression was not restricted to cap-dependent translation initiation, as both cap-dependent and IRES (internal ribosomal entry site)-dependent open reading frames of a bicistronic reporter gene were equally sensitive to the transfected miRNA. As pulse-labeling of nascent polypeptides indicated that repression occurred before completion of the synthesis of the full-length polypeptide chain, a ribosome drop-off model was proposed, in which miRNAs render ribosomes susceptible for premature translation termination.

Maroney and coworkers investigated the distribution of endogenous miRNAs and mRNAs in HeLa cells (Maroney et al. 2006). For instance, the KRAS mRNA, which is regulated by *let-7*, was found to be associated with translation competent ribosomes in the polysomal fractions. The finding that the KRAS mRNA remained in the polysomal fraction even under conditions known to interrupt translation initiation, argued against a ribosome drop-off and suggested a decelerating effect on the elongation rate.

However, a *let-7* mediated slow-down of the elongation rate in HeLa cells could not be observed for a reporter gene bearing the *C. elegans lin-41* 3′ UTR (Nottrott et al. 2006). Since the encoded protein remained undetectable, although reporter mRNA cosedimented with translation competent ribosomes, it was speculated that the nascent polypeptide was cotranslationally degraded. However, proteases involved in this process have not been identified, and in fact neither the inhibition of the proteasome nor the targeting of the reporter gene to the endoplasmic reticulum was found to restore protein accumulation in HeLa cells (Pillai et al. 2005). A model of cotranslational polypeptide degradation is thus based on negative evidence.

Cosedimentation of a considerable fraction of miRNAs or Argonaute proteins with polysomes was reported in many studies (Kim et al. 2004; Nelson et al. 2004; Nottrott et al. 2006; Olsen and Ambros 1999; Petersen et al. 2006). At first sight, this observation would argue against a mechanism that represses target gene translation initiation, as such a mechanism would deplete the miRNA target genes, and thus the miRNA and Argonaute, from the polysomal pool. However, a caveat to this interpretation is that efficient target gene repression might frequently require binding by several miRNAs (e.g., Doench et al. 2003; Vella et al. 2004). Thus, a substantial amount of miRNAs and Argonaute might be bound to polysomal mRNAs without greatly affecting translation.

2.5.2 Evidence for mRNA Deadenylation and Decay

Following a first report that showed that transfection of a miRNA into cultured cells resulted in reduced transcript levels for a number of apparently direct targets (Lim et al. 2005), nonendonucleolytic mRNA decay in response to miRNAs has been observed in *C. elegans* (Bagga et al. 2005) and many other systems. In zebrafish, *miR-430* was found to clear maternal mRNAs containing *miR-430* target sites at the onset of zygotic transcription (Giraldez et al. 2006). Depletion or ectopic expression of miRNAs alters the expression of validated miRNA targets or mRNAs containing binding sites for these miRNAs (Krutzfeldt et al. 2005; Lim et al. 2005; Linsley et al. 2007). Similarly, transcript levels of miRNA targets were found to increase in cells depleted of Dicer or Argonaute proteins (Rehwinkel et al. 2006; Schmitter et al. 2006).

MicroRNAs deploy the general mRNA degradation machinery to clear target mRNAs. Decapping and accelerated mRNA deadenylation have been observed in zebrafish, and fruitfly and human cell lines (Behm-Ansmant et al. 2006; Eulalio et al. 2007b; Giraldez et al. 2006; Wu et al. 2006). Target destabilization was found to depend on Argonaute proteins, the CAF1–CCR4–NOT deadenylase complex, the decapping enzyme DCP2, and the P-body component GW182 (Behm-Ansmant et al. 2006; Eulalio et al. 2007b). Depletion of these components leads to the stabilization of many miRNA target mRNAs that are otherwise degraded. Furthermore, Argonaute proteins, miRNAs, and repressed mRNAs are often found to colocalize to P-bodies, discrete cytoplasmic foci that harbor mRNA-catabolizing enzymes (Eulalio et al. 2007a).

Intriguingly, targets of *let-7* are destabilized to different degrees in different mammalian cell lines (Schmitter et al. 2006), and reporter mRNAs in *D. melanogaster* S2 cells can be silenced exclusively by either degradation or nondegradation, presumably translational repression or by a combination of both mechanisms (Behm-Ansmant et al. 2006; Eulalio et al. 2007b), suggesting that differences in cellular factors as well as the architecture or environment of miRNA target sites can influence the extent of target degradation.

How miRNAs initiate degradation of their target transcripts is not known. Moreover, it is unclear whether degradation is an independent mechanism or consequence of translational repression, as current evidence cannot distinguish between these two possibilities (Eulalio et al. 2008a; Filipowicz et al. 2008).

2.5.3 Evidence for Translational Repression at the Initiation Steps

Recent studies that recapitulated miRNA mediated gene repression in cell-free systems concluded that miRNAs interfere with target gene expression at translation initiation (Mathonnet et al. 2007; Thermann and Hentze 2007; Wakiyama et al. 2007; Wang et al. 2006). These studies unanimously reported a shift of repressed reporter genes to the monosomal pool of mRNA, consistent with reduced ribosome loading. This repression of translation initiation was found to depend on an m^7Gp-ppN-cap, whereas cap independent association of ribosomes via different IRES or ApppN-capped mRNAs was refractory to translational regulation. Inhibition of translation initiation has also been reported in cell-based approaches (Bhattacharyya et al. 2006; Pillai et al. 2005), which includes the only study explicitly showing this mechanism for an endogenous mRNA (Bhattacharyya et al. 2006).

Nevertheless, there is little agreement on the mechanisms that repress translation initiation. Human AGO2 (Argonaute 2) binds to a methylated cap analog in vitro via two tryptophan residues placed at an equivalent position in the initiation factor eIF4E (Kiriakidou et al. 2007). Thus, AGO2 miRNPs might compete with eIF4E for m⁷G-cap binding and thereby abrogate the bridging between m⁷G-cap and poly(A)-tail via eIF4G, which normally stimulates translation initiation. In line with disruption of mRNA circularization by eIF4F, whose subunits include eIF4E and eIF4G, eIF4F was found to be limiting for translational repression in mouse Krebs-2 cell extracts, and conversely, excess of eIF4F relieved translational repression (Mathonnet et al. 2007). Similarly, tethering of eIF4E and eIF4G to reporter constructs relieved translational repression in HeLa cells (Pillai et al. 2005). Nonetheless, recent work in fly cells suggests that cap-binding by AGO might not be sufficient to prevent translation initiation (Eulalio et al. 2008b).

If miRNAs repress translation initiation by interfering with mRNA circularization mediated by eIF4F, this would also imply a need for polyadenylation of the target transcript as a prerequisite for efficient circularization. However, the notion that a functional poly(A)-tail is necessary for translational regulation is controversial. Full miRNA mediated regulation of mRNA transfected into HeLa cells required a poly(A)-tail in one study (Humphreys et al. 2005), but not in another (Pillai et al. 2005). Moreover, in HEK293 cells, the poly(A) tail could be substituted by a histone stem-loop without eliminating repression (Eulalio et al. 2008b; Wu et al. 2006).

It has been suggested that translation initiation might be repressed by preventing 60S subunit joining, consistent with the finding that eIF6 was isolated in association with AGO2 and 60S ribosomes in HeLa cells (Chendrimada et al. 2007). eIF6 prevents premature assembly of the 60S and 40S ribosomal subunits by binding to 60S subunits. Recruitment of eIF6 by AGO2 could therefore interfere with translation initiation by preventing the recycling of ribosomal subunits. In *C. elegans*, RNAi against eIF6 led to an approximately twofold increase in LIN-14 and LIN-28 and their persistence at later time-points, when these proteins usually are not detected (Chendrimada et al. 2007). However, in our hands, depletion of eIF6 by RNAi induces slow growth, leaving it unclear whether the measured time-points indeed reflected two different developmental stages. Studies in mice, *D. melanogaster*, and *C. elegans* have indicated that eIF6 may not be generally required for miRNA function (Ding et al. 2008; Eulalio et al. 2007b, 2008b; Gandin et al. 2008) and it has been speculated that the involvement of eIF6 may be indirect, possibly reflecting a role in 60S subunit biogenesis (Filipowicz et al. 2008).

Although the precise mechanism and contributing factors remain unclear, various studies thus provide strong support for miRNA-mediated repression of translation initiation in vitro and ex vivo. Confusingly, however, this is precisely the mechanism that earlier studies in *C. elegans* appeared to rule out (Olsen and Ambros 1999; Seggerson et al. 2002). One possible conclusion is that miRNA functioned differently in *C. elegans* than in other organisms, or that indeed miRNAs studied in an intact organism, in vivo, behave differently from miRNAs studied in cultured cells or cell-free assays. The latter possibility is of particular concern given that almost all cell-based and cell-free studies have investigated transfected miRNA reporter genes, not endogenous target mRNAs, and both the modes of transfection (Lytle et al. 2007), and the promoter driving the reporter gene (Kong et al. 2008) have been reported to affect the apparent mode of miRNA-mediated gene repression. However, as we will discuss later, we have now demonstrated that miRNAs do indeed also repress translation initiation of their endogenous target mRNAs in *C. elegans* (Ding and Großhans 2009).

2.6   The *let-7* miRNA Extensively Interacts with Translation Factors

With the aim to study the interaction between *let-7* and the translation machinery under physiological conditions, we recently performed a reverse genetic screen (Ding et al. 2008). A major strength of *C. elegans* as a model organism is the simplicity of RNAi mediated knock-down of individual genes by feeding libraries of bacteria producing double-stranded RNA (Fraser et al. 2000; Kamath et al. 2003). The temperature sensitive *let-7(n2853)* allele harbors a point mutation in the mature *let-7* miRNA that impairs target mRNA silencing (Reinhart et al. 2000; Vella et al. 2004). As a consequence, mutant animals die by bursting through the vulva at the larval to adult transition when grown at 20°C or above. The lethality phenotype can be partially rescued by RNAi mediated knock-down of individual *let-7* target genes (Abrahante et al. 2003; Großhans et al. 2005; Lall et al. 2006; Lin et al. 2003; Slack et al. 2000). With the initial aim of identifying interaction partners of *let-7* in an unbiased approach, a library of 2,400 genes on chromosome I was screened for suppression of the *let-7* loss-of-function lethality phenotype. This initial screen identified 41 suppressors, including known and novel *let-7* target genes, as well as potential regulators of *let-7* expression, mediators of *let-7* activity and heterochronic genes (Ding et al. 2008). Twenty of these genes functioned in RNA or protein metabolism, among them several are putative subunits of eukaryotic translation initiation factors. When the screen was extended to include all translation factors with identifiable homologues in *C. elegans*, most of these, including initiation, elongation, and termination factors, partially suppressed the *let-7(n2853)* mutation.

Most *C. elegans* translation factors are thought to be essential, but RNAi typically achieves only partial depletion of targeted genes and animals were exposed to RNAi for only limited times. Larval development thus proceeded normally in most cases, although frequently slower than normal. To eliminate the possibility that this slow-growth contributed, indirectly, to suppression of *let-7(n2853)*-associated lethality, a subset of factors were depleted in wild-type animals, and shown to induce precocious differentiation of epidermal seam cells. This phenotype is consistent with a gain of *let-7* function, and suggests that suppression of *let-7* lethality is direct.

Unexpectedly, eIF6 was among the factors whose knock-down rescued *let-7(n2853)* animals and caused precocious seam cell differentiation in wild-type animals. In the light of the reported function of eIF6 as a mediator of *lin-4* function in *C. elegans* (Chendrimada et al. 2007), the opposite, *let-7* loss-of-function-like, retarded seam cell differentiation phenotype, would have been expected as a result of its depletion. However, recent studies on *D. melanogaster* S2 cells question a general role of eIF6 in promoting miRNA function (Eulalio et al. 2008b), and this might be reflected by our results.

C. elegans has readily recognizable orthologues of most of the translation factors commonly found in higher eukaryotes (Rhoads et al. 2006). Except for the termination factor eRF1, subunits of all translation factors were found to significantly suppress *let-7(n2853)* lethality. In addition to eIF6, we also examined the consequences of depleting eIF3 on seam cell differentiation, and again observed precocious differentiation in animals expressing functional *let-7*. eIF3 is required for the Met-tRNA$_i$ binding to the 40S ribosomal subunit and later for the recruitment of mRNA to the 43S pre-initiation complex (PIC) to form the 48S complex (Rhoads et al. 2006). The opposing roles of the tumor suppressor gene *let-7* and the eIF3 protooncogenes (Dong and Zhang 2006) are intriguing and may well be conserved beyond *C. elegans*: In humans, increased amounts of eIF3 stimulate translation of genes involved in cell proliferation (Zhang et al. 2007), for instance MYC and cyclin D1, which are also target genes of *let-7* (Bussing et al. 2008).

The eIF4 complex recruits the 43S PIC to mRNA. Depletion of its eIF4A subunit resulted in potent suppression, whereas depletion of eIF4G led to developmental arrest. No suppression was observed with eIF4E depletion, which at first sight is surprising, as many studies highlight the importance of m^{7}G-cap-binding for miRNA mediated translational regulation (Mathonnet et al. 2007; Thermann and Hentze 2007; Wakiyama et al. 2007; Wang et al. 2006), and it could be assumed that depletion of the cap-binding factor would favor the recently postulated cap-binding by AGO2 (Kiriakidou et al. 2007). However, the lack of an observable interaction is likely due to redundancy, as five different loci in the *C. elegans* genome encode eIF4E isoforms.

Taken together, these results pointed to a high sensitivity of *let-7* function to altered translation levels. Considering the studies supporting miRNA mediated translational control to occur after initiation in *C. elegans*, the identification of many translation initiation factors was somewhat surprising and prompted us to examine translational control on the mRNA level.

2.7 Polysome Profiling Confirms Translational Repression at the Initiation Step in *C. elegans*

We have recently reported that *let-7* represses translation initiation in *C. elegans*, demonstrating this mode of action for the first time in an organism (Ding and Großhans 2009). To assess whether *let-7* regulates translation initiation in vivo, we examined the polysome association of the two endogenous *let-7* target genes *daf-12* and *lin-41* in wild-type and *let-7(n2853)* animals, by applying whole animal lysates to sucrose density gradient centrifugation. In agreement with a decrease in translation initiation, *daf-12* and *lin-41* were moderately, but consistently, depleted from the highly translated polysomal fractions in wild-type animals (Fig. 2.3). However,

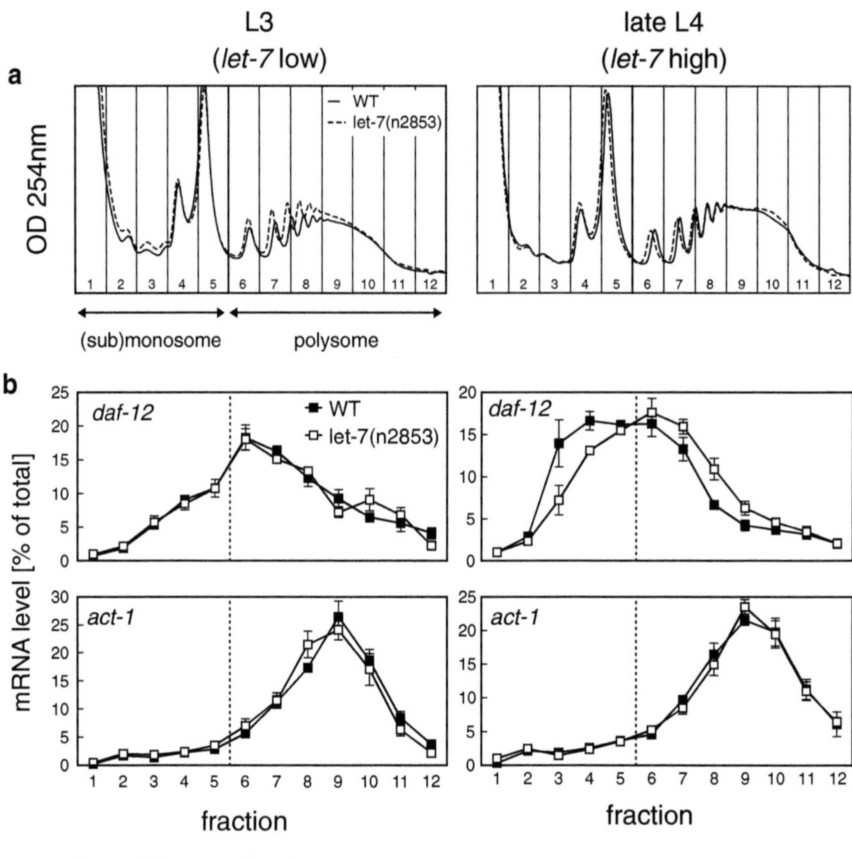

From Ding and Großhans, 2009

Fig. 2.3 *let-7* inhibits translation initiation of *daf-12* mRNA. (**a**) Polysome profiles of synchronized *wild-type* and *let-7(n2853)* animals at early L3, late L4. (**b**) Distribution of *daf-12* and *act-1* mRNA of across the fractions of the gradient. Before the onset of *let-7* expression in early L3, distribution of *daf-12* and *act-1* mRNA is essentially the same for wild-type and *let-7(n2853)* animals. In late L4, the distribution of the *let-7* target *daf-12* shifts to the (sub)-monosomal fractions in wild-type animals, whereas the distribution of *act-1*, which is not targeted by *let-7*, is not altered. Adapted from (Ding and Großhans 2009)

the limited degree of spatial and temporal coexpression of *let-7* miRNA and its targets limits the sensitivity of this assay. *let-7* is not universally expressed in *C. elegans* and as yet, regulation of target genes has been confirmed only in four different tissues, i.e., seam cells, ventral nerve cord, intestine, and head muscle (Abrahante et al. 2003; Großhans et al. 2005; Lall et al. 2006; Lin et al. 2003; Slack et al. 2000).

Although the heterogeneity of a whole animal system complicates the analysis, such a model has the benefit of providing a true physiological context. Improved sensitivity can then be obtained through tissue specific expression of miRNA target reporter genes. For instance, the apparent translational inhibition exerted by *let-7* considerably increased when a *lacZ* reporter gene carrying the *lin-41* 3′ UTR was directly expressed in epidermal seam cells, where *let-7* is also expressed (Fig. 2.4). Translational repression was specific, as translational repression of a *col-10::lacZ::lin-41* reporter gene relied on both wild-type *let-7* and the presence of previously described *let-7* binding sites (Vella et al. 2004).

Whereas previous reports on *lin-4* argued for translational repression downstream of initiation, the polysomal shifts observed in our experiments clearly demonstrated that *let-7* regulates two endogenous target genes by inhibiting translation at the initiation step. It thus appeared that two prominent miRNAs deployed two different modes of translational inhibition. To address this possibility, we examined the polysome association of transcripts in whole animal lysates of wild-type and *lin-4(e912)* mutant animals. In contrast to earlier studies (Olsen and Ambros 1999; Seggerson et al. 2002), we surprisingly discovered that *lin-4* also significantly inhibited translation initiation of its cognate target genes *lin-14* and

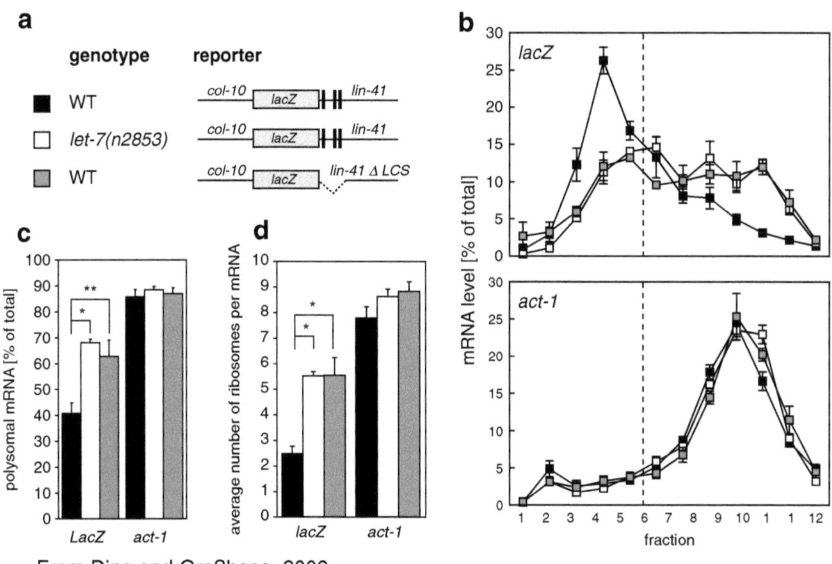

Fig. 2.4 Translational repression of *lin-41* is mediated by *let-7* and *let-7* binding sites. (**a**) Schematic representation of the reporter strains. The *lacZ* reporter genes were expressed in wild-type and *let-7(n2853)* animals under the control of the *col-10* promoter, which ensures constitutive expression in the seam cells, where *let-7* is also expressed. The vertical lines in the *lin-41* 3′ UTR represent *let-7* binding sites. In all experiments, synchronized late L4 animals were used. (**b**) Distribution of *lacZ* and *act-1* mRNA across the gradients. Only in the presence of both wild-type *let-7* and *let-7* binding sites, is the *lacZ* reporter gene translationally repressed. (**c**) Polysomal fraction of *lacZ*, endogenous *lin-41* and *act-1* as percentage of total RNA. (**d**) Average number of ribosomes on *lacZ* and *act-1* mRNA. (*, $p < 0.05$; **, $p < 0.01$; one-sided Student's *t*-test). Adapted from (Ding and Großhans 2009)

lin-28, indicating that *lin-4* and *let-7* function through the same mechanism. A cause for the discrepancy with the earlier data may be the fact that in earlier studies *lin-4* loss-of-function was approximated by comparing wild-type L1 animals to wild-type L2 animals as mature *lin-4* starts to accumulate at late L1 (Fig. 2.1). Thus, regulatory events occurring during *C. elegans* development, independently of *lin-4*, may have affected translational profiles of *lin-14* or *lin-28*.

In addition to translational repression, we also observed increased transcript levels of endogenous *daf-12* and *lin-41* mRNA in *let-7* mutant relative to wild-type animals (Ding and Großhans 2009), as previously observed with *lin-41* (Bagga et al. 2005). Transcript degradation might thus either provide an alternate mechanism for repression of miRNA target genes, or be a consequence of translational repression.

2.8 Inhibition of Translation Initiation and Transcript Degradation Both Depend on the GW182 Proteins AIN-1 and AIN-2

The *C. elegans* GW182 homolog AIN-1 (Argonaute interacting protein 1) has been identified through its function in developmental timing (Ding et al. 2005). The retarded heterochronic seam cell phenotype caused by *ain-1* loss-of-function mutations closely resembled the combined loss-of-function in the three *let-7* "sister" miRNAs *mir-48*, *mir-84*, and *mir-241*, which are related in sequence to *let-7* and function partially redundantly with it (Abbott et al. 2005; Lau et al. 2001; Lim et al. 2003). Genetic analysis of a reduction-of-function allele suggested that *ain-1* and its homolog *ain-2* function partially redundantly in post-transcriptional gene repression in *C. elegans* (Zhang et al. 2007). AIN-1 and AIN-2 were found to coimmunoprecipitate with DCR-1 (Dicer), mature miRNAs, and the Argonaute proteins ALG-1 and ALG-2, establishing GW182 proteins as bona fide components of the miRNA-induced silencing complex in *C. elegans* (Zhang et al. 2007). Complexes of GW182 proteins with Argonautes have also been identified in a variety of other organisms, including the human homologues TRNC6A-C (Behm-Ansmant et al. 2006; Eulalio et al. 2008b; Landthaler et al. 2008; Liu et al. 2005; Meister et al. 2005). Depletion of fly AGO1 or GW182 prevents the regulation of the same set of miRNA target genes, indicating that GW182 acts in the same pathway (Behm-Ansmant et al. 2006; Eulalio et al. 2007b). However, reporter genes mainly regulated at the translational level appeared less susceptible to GW182 depletion (Behm-Ansmant et al. 2006; Eulalio et al. 2007b), consistent with the proposed role of GW182 in directing miRNA targets to P-bodies for subsequent degradation (Ding et al. 2005). Nonetheless, the fact that miRNA target mRNAs could be coimmunoprecipitated with AIN-1/2 (Zhang et al. 2007) suggests that these mRNAs are at least partially stable under these conditions.

We attempted to uncouple translational repression and degradation by depleting the GW182 family members AIN-1 and AIN-2. To this end, we analyzed total

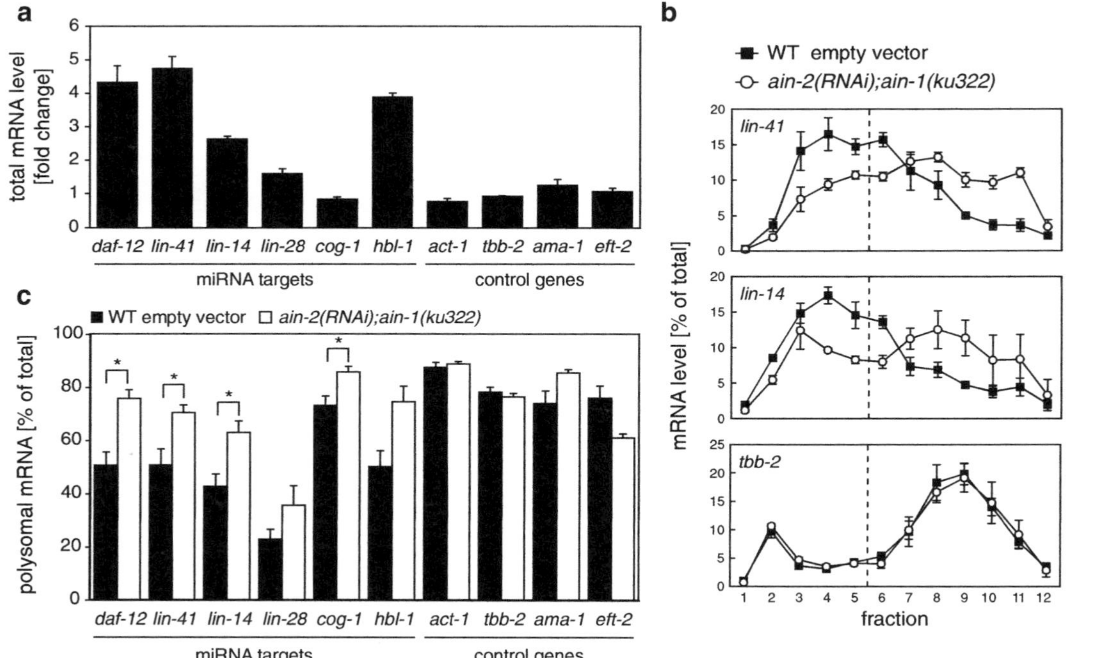

Fig. 2.5 The GW182 proteins AIN-1 and AIN-2 mediate translational repression and mRNA degradation. (**a**) Fold-change in mRNA levels between synchronized wild-type (empty vector RNAi control) animals and *ain-2(RNAi); ain-1(ku322)* animals at late L4 stage. mRNA levels were analyzed by quantitative reverse-transcription PCR and normalized to the average of the control genes *act-1, tbb-2, ama-1,* and *eft-2*. (**b**) Distribution of the *let-7* target *lin-41*, the *lin-4* target *lin-14* and, *tbb-2*, not known to be a miRNA target, across the gradient. (**c**) Polysomal fractions of miRNA target genes and control genes as percentage of the total mRNA. Synchronized late L4 wild-type (empty vector RNAi control) and *ain-2(RNAi); ain-1(ku322)* animals. (*, $p < 0.05$, *lin-28*: $p = 0.053$; *hbl-1*: $p = 0.056$; one-sided Student's *t*-test). Adapted from (Ding and Großhans 2009)

transcript levels and polysome profiles of wild-type and *ain-2(RNAi); ain-1(ku322)* double mutant animals. As anticipated, the combined depletion of AIN-1/2 resulted in a substantial increase in total *daf-12* and *lin-41* transcripts. To our surprise, however, the mutations also abrogated translational repression. In fact, the relief of translational repression caused by AIN-1/2 depletion exceeded that seen with the *let-7(n2853)* mutation, possibly reflecting residual *let-7* activity in *let-7(n2853)* animals and/or a redundant activity of the *let-7* family members *mir-48, mir-84,* and *mir-241*. We tested four additional miRNA target mRNAs (Fig. 2.5): *cog-1,* which is targeted by *lsy-6* in the ASEL head neuron (Johnston and Hobert 2003); *hbl-1,* which is targeted by *mir-48, mir-84, mir-241, let-7,* and *lin-4* (Abbott et al. 2005; Abrahante et al. 2003; Lin et al. 2003); and the *lin-4* targets *lin-14* and *lin-28* (Moss et al. 1997; Wightman et al. 1993). All of these showed the characteristic polysomal shifts in the *ain-2(RNAi); ain-1(ku322)* mutant relative to wild-type animals, confirming their translational repression by an AIN-1/-2-dependent mechanism. Of note, the total *cog-1* mRNA level remained unchanged, indicating that repression could also occur independently of target mRNA degradation. Consistent with our findings, a degradation independent, repressive function of GW182 has recently also been shown with miRNA target reporter genes in Drosophila cells (Eulalio et al. 2008b). Taken together, our results demonstrate that repression of translation initiation by miRNAs is wide-spread in *C. elegans* and requires AIN-1/2.

2.9   Conclusions and Future Perspectives

Experiments on miRNA modes of action in vitro, ex vivo, and in vivo have previously yielded disparate results. The first two approaches predominantly, although not exclusively, supported repression of translation initiation and transcript degradation. By contrast, in vivo studies yielded conflicting results on the relevance of degradation and appeared to rule out repression of translation initiation. It was possible that these disparities reflected true mechanistic differences in different organisms, consistent with the fact that the in vivo work largely relied on *C. elegans,* whereas the other two approaches utilized human and Drosophila cells. More disconcertingly, ex vivo and in vitro studies had almost exclusively relied on transfected miRNA target reporter genes and two studies raised concerns that the transfection procedures and the promoters used to express these reporter genes influenced the apparent mode of miRNA activity. Our recent work now demonstrates that repression of translation initiation by miRNAs also occurs in vivo, in *C. elegans,* and on endogenous mRNAs targeted by three different miRNAs. Thus, miRNAs have now been shown to mediate repression of translation initiation in vivo, ex vivo, and in vitro, on both endogenous targets and reporter mRNAs, making a particularly compelling case for this mode of repression.

Loss-of-function of the GW182 homologues AIN-1 and AIN-2 relieves miRNA-mediated gene repression, supporting the notion that these proteins are essential miRNA effectors in *C. elegans,* consistent with the developmental defects observed in earlier studies. Although AIN-1/-2 are required for both translational repression and transcript degradation, it is unclear whether these two constitute independent mechanisms or whether target degradation is a consequence of translational repression. However, at least for the *lsy-6* target *cog-1*, translational repression is not accompanied by target degradation, and we do not observe a correlation between the extent of translational repression and target gene degradation for various other miRNA:target pairs that we tested, which may hint at two distinct mechanisms. AIN-1 and AIN-2 may then coordinate translational repression and target degradation, possibly by interacting with distinct mediators or effectors. Future work directed towards the identification of these mediators and effectors may solve the question whether translational control and target degradation are a result of functionally distinct silencing complexes, and therefore, may be uncoupled. Now that both mechanisms have been demonstrated in *C. elegans*, its powerful genetic tools can be brought to bear on the issue. Detailed dissection of the genetic interaction partners of *let-7* that we recently uncovered might provide an avenue into identifying the factors involved.

Acknowledgments X.D. was supported by a PhD student fellowship from the Boehringer Ingelheim Foundation. Research in H.G.'s lab is funded by the Novartis Research Foundation and the Swiss National Science Foundation.

References

Abbott AL, Alvarez-Saavedra E, Miska EA, Lau NC, Bartel DP, Horvitz HR, Ambros V (2005) The let-7 MicroRNA family members mir-48, mir-84, and mir-241 function together to regulate developmental timing in Caenorhabditis elegans. Dev Cell 9:403–414

Abrahante JE, Daul AL, Li M, Volk ML, Tennessen JM, Miller EA, Rougvie AE (2003) The Caenorhabditis elegans hunchback-like gene lin-57/hbl-1 controls developmental time and is regulated by microRNAs. Dev Cell 4:625–637

Bagga S, Bracht J, Hunter S, Massirer K, Holtz J, Eachus R, Pasquinelli AE (2005) Regulation by let-7 and lin-4 miRNAs results in target mRNA degradation. Cell 122:553–563

Behm-Ansmant I, Rehwinkel J, Doerks T, Stark A, Bork P, Izaurralde E (2006) mRNA degradation by miRNAs and GW182 requires both CCR4:NOT deadenylase and DCP1:DCP2 decapping complexes. Genes Dev 20:1885–1898

Bhattacharyya SN, Habermacher R, Martine U, Closs EI, Filipowicz W (2006) Relief of microRNA-mediated translational repression in human cells subjected to stress. Cell 125:1111–1124

Bussing I, Slack FJ, Großhans H (2008) let-7 microRNAs in development, stem cells and cancer. Trends Mol Med 14:400–409

Chalfie M, Horvitz HR, Sulston JE (1981) Mutations that lead to reiterations in the cell lineages of C. elegans. Cell 24:59–69

Chendrimada TP, Finn KJ, Ji X, Baillat D, Gregory RI, Liebhaber SA, Pasquinelli AE, Shiekhattar R (2007) MicroRNA silencing through RISC recruitment of eIF6. Nature 447:823–828

Denli AM, Tops BB, Plasterk RH, Ketting RF, Hannon GJ (2004) Processing of primary microRNAs by the Microprocessor complex. Nature 432:231–235

Ding XC, Großhans H (2009) Repression of C. elegans microRNA targets at the initiation level of translation requires GW182 proteins. EMBO J 28:213–222

Ding L, Spencer A, Morita K, Han M (2005) The developmental timing regulator AIN-1 interacts with miRISCs and may target the argonaute protein ALG-1 to cytoplasmic P bodies in C. elegans. Mol Cell 19:437–447

Ding XC, Slack FJ, Großhans H (2008) The let-7 microRNA interfaces extensively with the translation machinery to regulate cell differentiation. Cell Cycle 7:3083–3090

Doench JG, Petersen CP, Sharp PA (2003) siRNAs can function as miRNAs. Genes Dev 17:438–442

Dong Z, Zhang JT (2006) Initiation factor eIF3 and regulation of mRNA translation, cell growth, and cancer. Crit Rev Oncol Hematol 59:169–180

Eulalio A, Behm-Ansmant I, Izaurralde E (2007a) P bodies: at the crossroads of post-transcriptional pathways. Nat Rev Mol Cell Biol 8:9–22

Eulalio A, Rehwinkel J, Stricker M, Huntzinger E, Yang SF, Doerks T, Dorner S, Bork P, Boutros M, Izaurralde E (2007b) Target-specific requirements for enhancers of decapping in miRNA-mediated gene silencing. Genes Dev 21:2558–2570

Eulalio A, Huntzinger E, Izaurralde E (2008a) Getting to the root of miRNA-mediated gene silencing. Cell 132:9–14

Eulalio A, Huntzinger E, Izaurralde E (2008b) GW182 interaction with Argonaute is essential for miRNA-mediated translational repression and mRNA decay. Nat Struct Mol Biol 15:346–353

Filipowicz W, Bhattacharyya SN, Sonenberg N (2008) Mechanisms of post-transcriptional regulation by microRNAs: are the answers in sight? Nat Rev Genet 9:102–114

Fraser AG, Kamath RS, Zipperlen P, Martinez-Campos M, Sohrmann M, Ahringer J (2000) Functional genomic analysis of C. elegans chromosome I by systematic RNA interference. Nature 408:325–330

Gandin V, Miluzio A, Barbieri AM, Beugnet A, Kiyokawa H, Marchisio PC, Biffo S (2008) Eukaryotic initiation factor 6 is rate-limiting in translation, growth and transformation. Nature 455:684–688

Giraldez AJ, Mishima Y, Rihel J, Grocock RJ, Van Dongen S, Inoue K, Enright AJ, Schier AF (2006) Zebrafish MiR-430 promotes deadenylation and clearance of maternal mRNAs. Science 312:75–79

Grishok A, Pasquinelli AE, Conte D, Li N, Parrish S, Ha I, Baillie DL, Fire A, Ruvkun G, Mello CC (2001) Genes and mechanisms related to RNA interference regulate expression of the small temporal RNAs that control C. elegans developmental timing. Cell 106:23–34

Großhans H, Slack FJ (2002) Micro-RNAs: small is plentiful. J Cell Biol 156:17–21

Großhans H, Johnson T, Reinert KL, Gerstein M, Slack FJ (2005) The temporal patterning microRNA let-7 regulates several transcription factors at the larval to adult transition in C. elegans. Dev Cell 8:321–330

Horvitz HR, Sulston JE (1980) Isolation and genetic characterization of cell-lineage mutants of the nematode Caenorhabditis elegans. Genetics 96:435–454

Humphreys DT, Westman BJ, Martin DI, Preiss T (2005) MicroRNAs control translation initiation by inhibiting eukaryotic initiation factor 4E/cap and poly(A) tail function. Proc Natl Acad Sci U S A 102:16961–16966

Johnston RJ, Hobert O (2003) A microRNA controlling left/right neuronal asymmetry in Caenorhabditis elegans. Nature 426:845–849

Kamath RS, Fraser AG, Dong Y, Poulin G, Durbin R, Gotta M, Kanapin A, Le Bot N, Moreno S, Sohrmann M et al (2003) Systematic functional analysis of the Caenorhabditis elegans genome using RNAi. Nature 421:231–237

Ketting RF, Fischer SE, Bernstein E, Sijen T, Hannon GJ, Plasterk RH (2001) Dicer functions in RNA interference and in synthesis of small RNA involved in developmental timing in C. elegans. Genes Dev 15:2654–2659

Kim J, Krichevsky A, Grad Y, Hayes GD, Kosik KS, Church GM, Ruvkun G (2004) Identification of many microRNAs that copurify with polyribosomes in mammalian neurons. Proc Natl Acad Sci U S A 101:360–365

Kiriakidou M, Tan GS, Lamprinaki S, De Planell-Saguer M, Nelson PT, Mourelatos Z (2007) An mRNA m(7)G cap binding-like motif within human Ago2 represses translation. Cell 129:1141–1151

Knight SW, Bass BL (2001) A role for the RNase III enzyme DCR-1 in RNA interference and germ line development in Caenorhabditis elegans. Science 293:2269–2271

Kong YW, Cannell IG, de Moor CH, Hill K, Garside PG, Hamilton TL, Meijer HA, Dobbyn HC, Stoneley M, Spriggs KA et al (2008) The mechanism of micro-RNA-mediated translation repression is determined by the promoter of the target gene. Proc Natl Acad Sci U S A 105:8866–8871

Krutzfeldt J, Rajewsky N, Braich R, Rajeev KG, Tuschl T, Manoharan M, Stoffel M (2005) Silencing of microRNAs in vivo with 'antagomirs'. Nature 438:685–689

Lall S, Grun D, Krek A, Chen K, Wang YL, Dewey CN, Sood P, Colombo T, Bray N, Macmenamin P et al (2006) A genome-wide map of conserved microRNA targets in C. elegans. Curr Biol 16:460–471

Landthaler M, Gaidatzis D, Rothballer A, Chen PY, Soll SJ, Dinic L, Ojo T, Hafner M, Zavolan M, Tuschl T (2008) Molecular characterization of human Argonaute-containing ribonucleoprotein complexes and their bound target mRNAs. RNA 14:2580–2596

Lau NC, Lim LP, Weinstein EG, Bartel DP (2001) An abundant class of tiny RNAs with probable regulatory roles in Caenorhabditis elegans. Science 294:858–862

Lee RC, Feinbaum RL, Ambros V (1993) The C. elegans heterochronic gene lin-4 encodes small RNAs with antisense complementarity to lin-14. Cell 75:843–854

Lim LP, Lau NC, Weinstein EG, Abdelhakim A, Yekta S, Rhoades MW, Burge CB, Bartel DP (2003) The microRNAs of Caenorhabditis elegans. Genes Dev 17:991–1008

Lim LP, Lau NC, Garrett-Engele P, Grimson A, Schelter JM, Castle J, Bartel DP, Linsley PS, Johnson JM (2005) Microarray analysis shows that some microRNAs downregulate large numbers of target mRNAs. Nature 433:769–773

Lin SY, Johnson SM, Abraham M, Vella MC, Pasquinelli A, Gamberi C, Gottlieb E, Slack FJ (2003) The C elegans hunchback homolog, hbl-1, controls temporal patterning and is a probable microRNA target. Dev Cell 4:639–650

Linsley PS, Schelter J, Burchard J, Kibukawa M, Martin MM, Bartz SR, Johnson JM, Cummins JM, Raymond CK, Dai H et al (2007) Transcripts targeted by the microRNA-16 family cooperatively regulate cell cycle progression. Mol Cell Biol 27:2240–2252

Liu J, Valencia-Sanchez MA, Hannon GJ, Parker R (2005) MicroRNA-dependent localization of targeted mRNAs to mammalian P-bodies. Nat Cell Biol 7:719–723

Lytle JR, Yario TA, Steitz JA (2007) Target mRNAs are repressed as efficiently by microRNA-binding sites in the 5' UTR as in the 3' UTR. Proc Natl Acad Sci U S A 104:9667–9672

Maroney PA, Yu Y, Fisher J, Nilsen TW (2006) Evidence that microRNAs are associated with translating messenger RNAs in human cells. Nat Struct Mol Biol 13:1102–1107

Mathonnet G, Fabian MR, Svitkin YV, Parsyan A, Huck L, Murata T, Biffo S, Merrick WC, Darzynkiewicz E, Pillai RS et al (2007) MicroRNA inhibition of translation initiation in vitro by targeting the cap-binding complex eIF4F. Science 317:1764–1767

Meister G, Landthaler M, Peters L, Chen PY, Urlaub H, Luhrmann R, Tuschl T (2005) Identification of novel argonaute-associated proteins. Curr Biol 15:2149–2155

Moss EG (2007) Heterochronic genes and the nature of developmental time. Curr Biol 17:R425–R434

Moss EG, Lee RC, Ambros V (1997) The cold shock domain protein LIN-28 controls developmental timing in C. elegans and is regulated by the lin-4 RNA. Cell 88:637–646

Nelson PT, Hatzigeorgiou AG, Mourelatos Z (2004) miRNP:mRNA association in polyribosomes in a human neuronal cell line. RNA 10:387–394

Nottrott S, Simard MJ, Richter JD (2006) Human let-7a miRNA blocks protein production on actively translating polyribosomes. Nat Struct Mol Biol 13:1108–1114

Olsen PH, Ambros V (1999) The lin-4 regulatory RNA controls developmental timing in Caenorhabditis elegans by blocking LIN-14 protein synthesis after the initiation of translation. Dev Biol 216:671–680

Pasquinelli AE, Reinhart BJ, Slack F, Martindale MQ, Kuroda MI, Maller B, Hayward DC, Ball EE, Degnan B, Muller P et al (2000) Conservation of the sequence and temporal expression of let-7 heterochronic regulatory RNA. Nature 408:86–89

Petersen CP, Bordeleau ME, Pelletier J, Sharp PA (2006) Short RNAs repress translation after initiation in mammalian cells. Mol Cell 21:533–542

Pillai RS, Bhattacharyya SN, Artus CG, Zoller T, Cougot N, Basyuk E, Bertrand E, Filipowicz W (2005) Inhibition of translational initiation by Let-7 MicroRNA in human cells. Science 309:1573–1576

Rehwinkel J, Natalin P, Stark A, Brennecke J, Cohen SM, Izaurralde E (2006) Genome-wide analysis of mRNAs regulated by Drosha and Argonaute proteins in Drosophila melanogaster. Mol Cell Biol 26:2965–2975

Reinhart BJ, Slack FJ, Basson M, Pasquinelli AE, Bettinger JC, Rougvie AE, Horvitz HR, Ruvkun G (2000) The 21-nucleotide let-7 RNA regulates developmental timing in Caenorhabditis elegans. Nature 403:901–906

Rhoads RE, Dinkova TD, Korneeva NL (2006) Mechanism and regulation of translation in C. elegans. WormBook 1–18

Schmitter D, Filkowski J, Sewer A, Pillai RS, Oakeley EJ, Zavolan M, Svoboda P, Filipowicz W (2006) Effects of Dicer and Argonaute down-regulation on mRNA levels in human HEK293 cells. Nucl Acids Res 34:4801–4815

Seggerson K, Tang L, Moss EG (2002) Two genetic circuits repress the Caenorhabditis elegans heterochronic gene lin-28 after translation initiation. Dev Biol 243:215–225

Slack FJ, Basson M, Liu Z, Ambros V, Horvitz HR, Ruvkun G (2000) The lin-41 RBCC gene acts in the C. elegans heterochronic pathway between the let-7 regulatory RNA and the LIN-29 transcription factor. Mol Cell 5:659–669

Thermann R, Hentze MW (2007) Drosophila miR2 induces pseudo-polysomes and inhibits translation initiation. Nature 447:875–878

Vella MC, Choi EY, Lin SY, Reinert K, Slack FJ (2004) The C. elegans microRNA let-7 binds to imperfect let-7 complementary sites from the lin-41 3'UTR. Genes Dev 18:132–137

Wakiyama M, Takimoto K, Ohara O, Yokoyama S (2007) Let-7 microRNA-mediated mRNA deadenylation and translational repression in a mammalian cell-free system. Genes Dev 21:1857–1862

Wang B, Love TM, Call ME, Doench JG, Novina CD (2006) Recapitulation of short RNA-directed translational gene silencing in vitro. Mol Cell 22:553–560

Wettstein FO, Staehelin T, Noll H (1963) Ribosomal aggregate engaged in protein synthesis: characterization of the ergosome. Nature 197:430–435

Wightman B, Ha I, Ruvkun G (1993) Posttranscriptional regulation of the heterochronic gene lin-14 by lin-4 mediates temporal pattern formation in C. elegans. Cell 75:855–862

Wu L, Belasco JG (2005) Micro-RNA regulation of the mammalian lin-28 gene during neuronal differentiation of embryonal carcinoma cells. Mol Cell Biol 25:9198–9208

Wu L, Fan J, Belasco JG (2006) MicroRNAs direct rapid deadenylation of mRNA. Proc Natl Acad Sci U S A 103:4034–4039

Zhang L, Ding L, Cheung TH, Dong MQ, Chen J, Sewell AK, Liu X, Yates JR 3rd, Han M (2007) Systematic identification of C. elegans miRISC proteins, miRNAs, and mRNA targets by their interactions with GW182 proteins AIN-1 and AIN-2. Mol Cell 28:598–613

Chapter 3
Translational Inhibition by MicroRNAs in Plants

Bin Yu and Hai Wang

Contents

Abstract MicroRNAs (miRNAs) are 21–24 nucleotide riboregulators, which selectively repress gene expression through transcript cleavage and/or translational inhibition. It was thought that most plant miRNAs act through target transcript cleavage due to the high degree of complementarity between miRNAs and their targets. However, recent studies have suggested widespread translational inhibition by miRNAs in plants. The mechanisms underlining translational inhibition by plant miRNAs are largely unknown, but existing evidence has indicated that plants and animals share some mechanistic similarity of translational inhibition. Translational inhibition by miRNAs has been shown to regulate floral patterning, floral timing, and stress responses. This chapter covers recent progress on plant miRNA-mediated translational control.

B. Yu (✉) and H. Wang
School of Biological Sciences & Center for Plant Science Innovation, University of Nebraska,
Lincoln, NE 68588, USA
e-mail: byu3@unl.edu

R.E. Rhoads (ed.), *miRNA Regulation of the Translational Machinery*,   41
Progress in Molecular and Subcellular Biology 50,
DOI 10.1007/978-3-642-03103-8_3, © Springer-Verlag Berlin Heidelberg 2010

3.1   Introduction

MicroRNAs (miRNAs) are 21–24 nucleotide noncoding RNAs that inhibit the expression of genes containing partially complementary sequences at post-transcriptional levels through translational inhibition or target cleavage (Bartel 2004), or sometimes at the transcriptional level through chromatin modification (Bao et al. 2004). Plant miRNAs were first identified in 2002 (Llave et al. 2002a; Mette et al. 2002; Park et al. 2002; Reinhart et al. 2002), approximately a decade after the shorter lin-4 RNA, the founding member of miRNAs, was identified in *Caenortabditis elegans* (Lee et al. 1993). Since then, the functions of plant miRNAs in regulating biological processes including development, metabolism, hormone responses, responses to biotic and abiotic stress, and others have been established (Mallory and Vaucheret 2006). Owing to the development of computational prediction algorithms and large-scale sequence techniques, hundreds of miRNAs have been identified in plants (Meyers et al. 2006). The great potential of these miRNAs to regulate thousands of genes has emphasized the importance of a riboregulatory network of gene expression in addition to transcriptional factors.

Although translational inhibition by miRNAs has been reported in plants (Aukerman and Sakai 2003; Chen 2004; Dugas and Bartel 2008; Gandikota et al. 2007), target-cleavage by miRNAs is thought to be the predominant way for miRNAs to act (Bartel 2004). However, recent studies have shown that translational inhibition is widely present in plants (Brodersen et al. 2008). In addition to miRNAs, plants are also enriched with small interfering RNAs (siRNAs). Translational inhibition by siRNAs in plants has been demonstrated recently (Brodersen et al. 2008). This chapter summarizes the recent knowledge about the biogenesis of plant miRNAs and siRNAs, miRNA-mediated translational control in plants, and translational inhibition by siRNAs.

3.2   miRNA and siRNA Biogenesis

3.2.1   *Plant miRNA Biogenesis*

miRNAs are generated from long transcripts by DNA-dependent RNA polymerase II (pol II) (Bartel 2004). The primary transcripts of miRNA genes, containing a hairpin structure, carry a seven-methyl guanosine (m7G) cap at the 5′ end and a polyadenosine tail (polyA) at the 3′ end (Bartel 2004). Plant miRNA genes are independent transcription units subjected to transcriptional regulation with a few exceptions (Rajagopalan et al. 2006). For instance, miR838 are derived from intron 14 of *DICER-LIKE1* gene (Rajagopalan et al. 2006).

In Arabidopsis and Rice, an RNAase III like domain-containing protein, called Dicer-like 1 (DCL1), releases a single miRNA/miRNA* duplex from the hairpin structure of pri-miRNAs through two-step cleavages in the nucleus excluding miR822 and miR839, which are generated by DCL4 (Rajagopalan et al. 2006; Ramachandran and Chen 2008b). The miRNA/miRNA* has a 2-nt overhang at the 3′ end of each strand and a phosphate group at the 5′ end of each strand, which are typical features of RNAase III cleavage products (Bartel 2004). The efficient and accurate processing of pri-miRNAs requires a double-strand RNA binding protein HYPONASTIC LEAVES1 (HYL1) and a zinc finger protein SERRATE (SE) (Dong et al. 2008; Han et al. 2004; Lobbes et al. 2006; Reinhart et al. 2002; Yang et al. 2006a). DCL1, HYL1 and SE form a small nuclear body containing pri-miRNAs, called D-body or SmD3/SmB nuclear bodies (Fang and Spector 2007; Kurihara and Watanabe 2004; Song et al. 2007).

It was reported recently that the loss-of-function of ABH1/CBP80 and CBP20, two subunits of nuclear cap-binding complex, reduces the levels of miRNAs, suggesting the involvement of the multifunctional cap-binding complex in pri-miRNA processing (Chen, 2008; Gregory et al. 2008; Laubinger et al. 2008). In addition, it was proposed that DAWDLE (DDL), a forkhead domain containing protein, might participate in the miRNA biogenesis by stabilizing pri-miRNAs and facilitating their access or recognition by DCL1 (Yu et al. 2008). It has been shown that DDL interacts with DCL1, and lack of DDL reduces the levels of pri-miRNAs, pre-miRNAs, and mature miRNAs (Yu et al. 2008). Interestingly, Smad interacting protein1 (SNIP1), a human ortholog of DDL, interacts with Drosha and participates in the miRNA biogenesis, suggesting that DDL is an evolutionarily conserved factor in the miRNA biogenesis (Yu et al. 2008).

After generation, the miRNA/miRNA* duplexes are methylated on the 2′OH of the 3′ terminal ribose on each strand by a protein named HUA EHANCER1 (HEN1) (Yang et al. 2006b; Yu et al. 2005). Currently, it is not clear whether the methylation occurs in the nucleus or cytoplasm. In plants carrying a *hen1* mutation, lack of miRNA methylation reduces miRNA abundance and causes the addition of 1–6 uridines at miRNA 3′ terminal, suggesting that miRNA methylation protects miR-NAs from degradation and/or uridylation (Li et al. 2005). The genes encoding these enzymes are unknown. A recent study showed that a family of exoribonucleases encoded by the SMALL RNA DEGRADING NUCLEASE (SDN) genes degrades single-stranded mature miRNAs in Arabidopsis (Ramachandran and Chen 2008a). In Arabidopsis, an ortholog of exportin-5 called HASTY exports mature miRNAs to the cytoplasm, although it is not clear whether miRNAs are exported as duplex or single strand (Park et al. 2005). However, mutations in HASTY do not reduce the levels of several miRNAs, suggesting the presence of an alternative exporting mechanisms (Park et al. 2005).

3.2.2   SiRNA Biogenesis

Beyond miRNAs, plants are also enriched with siRNAs, which represent 85% of cellular small RNAs (Kasschau et al. 2007; Lu et al. 2006; Rajagopalan et al. 2006; Zhang et al. 2007). The difference between miRNAs and siRNAs is their origin (Chen 2005). While miRNA is excised from a partially complementary hairpin structure in the pri-miRNA, siRNA is produced from a hairpin transgene or a perfect complementary long dsRNA converted by RNA dependent polymerases (RDRs) from a single-stranded RNA, or resulted from sense and antisense transcription (Chen 2005).

A class of 24 nt RNAs produced from transposons and repetitive DNA consists the largest portion (84%) of small RNAs. The function of this class siRNAs includes directing DNA methylation and histon modification (Mallory and Vaucheret 2006). These siRNAs are excised by DCL3, a homolog of DCL1, from dsRNAs, which are presumably converted by RNA dependent RNA polymerase 2 from transcripts of repeat DNAs (Xie et al. 2004). The biogenesis of these siRNAs requires the plant specific DNA-dependent polymerase IV, Pol IVa (Herr et al. 2005; Kanno et al. 2005; Onodera et al. 2005). Pol IVb, another form of Pol IV is required for the function of these siRNAs and is involved in the biogenesis of some of them (Kanno et al. 2005; Pontier et al. 2005).

Trans-acting siRNAs (ta-siRNAs), consisting ~1% of small RNA, represent another class of siRNAs (Allen et al. 2005; Vazquez et al. 2004; Yoshikawa et al. 2005). ta-siRNAs are generated from noncoding transcripts. The transcripts are first subjected to miRNA-mediated cleavage. The cleavage fragments are then stabilized by SGS3 and converted to dsRNAs by RDR6, which are processed by DCL4 into 21 nt siRNAs (Allen et al. 2005; Vazquez et al. 2004; Yoshikawa et al. 2005). The process of ta-siRNA biogenesis also requires SDE5, whose function is unknown, and DOUBLE STRAND RNA BINDING PROTEIN4 (DRB4), which is a homology of HYL1. ta-siRNAs act on genes other than their originating genes (Adenot et al. 2006; Hernandez-Pinzon et al. 2007).

The third class of siRNAs is natural sense–antisense siRNAs (Nat-siRNAs) identified under biotic or abiotic stresses (Borsani et al. 2005; Katiyar-Agarwal et al. 2006). They are generated from bidirectional transcripts. RDR6, SGS3, Pol IVa, HYL1, HEN1, and two DCL proteins are involved in the biogenesis of *nat-siRNAs*. Recently, a class of 30–40nt long siRNAs (lsiRNAs) was isolated under biotic stress or specific growth conditions (Katiyar-Agarwal et al. 2007). The biogenesis of this class siRNAs is dependent on HYL1, HST, RDR6, and Pol IV (Katiyar-Agarwal et al. 2007).

3.3 ARGONAUTE Proteins

Through base-pairing with the complementary sequence embedded in the targets, small RNAs guide transcriptional and post-transcriptional silencing performed by a ribonucleoprotein complex called RNA induced silencing complex (RISC) (Bartel 2004). Members of Argonaute (AGO) protein family are the effectors in the RISC complex (Vaucheret 2008). *AGO1* is the founding member of this gene family identified from a genetic screen for mutants deficient in plant growth (Bohmert et al. 1998; Vaucheret 2008). It was named because of the resemblance of appearance between *ago1* and a small squid of *Agrounauta* genus (Bohmert et al. 1998). AGO proteins are conserved among eukaryotes. They contain conserved PAZ, MID, and

PIWI domains in the C-termini. It has been shown that the MID domain associates with the 5′ phosphate of small RNAs, the PAZ domain binds to the 3′ terminal nucleotide, and the PIWI domain, which contains an RNaseH signature, exhibits endonuclease activity and performs target cleavage.

The numbers of AGO proteins are diversified among different plant species. While unicellular green algae *Chlamydomonas reinhardtii* encodes only two AGOs (Zhao et al. 2007b), Arabidopsis and rice contain 10 and 18 AGOs, respectively (Morel et al. 2002; Nonomura et al. 2007). Among ten AGOs of Arabidopsis, AGO1, AGO7, and AGO4 have the slicing activity. It was found that AGO1 is responsible for most miRNA-mediated mRNA cleavage (Baumberger and Baulcombe 2005), and AGO7 is the slicer of miR390-mediated target cleavage (Montgomery et al. 2008). AGO4 acts redundantly with AGO6 in siRNA-mediated DNA methylation of histon modification (Zheng et al. 2007).

The AGO proteins selectively recruit the miRNA strand with a less stable 5′ end from the miRNA/miRNA* duplex into RISC complex. The other strand named miRNA* is then degraded. Intriguingly, deep sequence analysis of small RNAs coimmunoprecipitated with AGO proteins revealed that small RNAs are selectively channeled into RISC complexes according to their 5′ terminal nucleotide (Mi et al. 2008; Montgomery et al. 2008; Takeda et al. 2008). AGO1 preferentially binds to miRNAs with a 5′ uridine (Mi et al. 2008). AGO2 and AGO4 complexes harbor small RNAs with a 5′ adenine, and AGO5 associate with small RNAs with a 5′ cytosine (Mi et al. 2008). Substituting the 5′ terminal uridine of miR391 or miR171, which are associated with AGO1, with an adenine forces them into an AGO2 complex, indicating an essential role of 5′ terminal nucleotides in sorting small RNAs into RISC complex (Mi et al. 2008). However, there are a few exceptions. For instance, miR172 and miR390 with a 5′ terminal A are preferentially associated with AGO1 and AGO7, respectively (Mi et al. 2008; Montgomery et al. 2008).

In addition, sequence analysis showed that AGO4 associates with miR172 in vivo, and the AGO4 complex harboring miR172 cleaves miRNA target in vitro (Qi et al. 2006). However, plants lacking of AGO4 do not show miR172-resistant phenotypes, indicating that AGO4 is not required for miR172 function in vivo (Qi et al. 2006). This result raises question about the functional role of AGO4–miR172 association in Arabidopsis (Qi et al. 2006; Vaucheret 2008).

3.4 Translational Inhibition by Small RNAs is Common in Plants

3.4.1 Translational Inhibition by miRNAs

miRNAs regulate gene expression through translational inhibition and/or target cleavage. A key factor determining the functional mechanism of miRNAs is the complementarity between miRNAs and their targets (Bartel 2004). It was observed that a perfect match between miRNAs and their targets promotes target cleavage, while central mismatch inhibits cleavage and enables translational inhibition (Bartel 2004). Most animal miRNAs have imperfect matches with their targets and repress the translation, while plant miRNAs are highly complementary to their targets and were thought to function predominantly by target-cleavage (Bartel 2004).

However, it was observed recently that translational inhibition by miRNA is a widespread phenomenon (Brodersen et al. 2008). In a genetic screen for Arabidopsis mutants deficient in repressing the expression of a green fluorescent protein (GFP) containing a miR171 target site immediately downstream of the stop codon, two *microRNA action deficient (mad)* mutants, *mad5 and mad6*, in which the levels of GFP protein but not mRNA are increased were isolated (Brodersen et al. 2008). Further analysis of several targets of endogenous miRNAs showed that the protein levels of all these targets are increased in *mad5* and *mad6* compared with those in WT (Brodersen et al. 2008). In contrast, the mRNA levels of most targets remain unchanged (Brodersen et al. 2008). These miRNA targets represent the distribution of target site in the 5′ UTR, coding sequence, or 3′ UTR of mRNAs and have different degree of complementarity with miRNAs (Brodersen et al. 2008). These results suggested that translational inhibition is a common action mechanism of plant miRNAs regardless of the localization of target site in the target and the degree of complementarity.

3.4.2 *Translational Inhibition by siRNAs*

Translational inhibition by siRNAs has been established in animals but not in plants. The SUC–SUL (SS) silencing system represses the expression of the endogenous SULFUR mRNA (SUL) in Arabidopsis through the action of DCL4 dependent siRNAs, which are generated from phloem-specific expression of an inverted-repeat (IR). Silencing of SUL results in a vein-centered chlorotic pheno-type. Silencing of SUL requires AGO1 protein. Introducing an *ago1-27* mutation into the SUL silencing line represses the silencing without changing the levels of SUL mRNA and siRNAs. In addition, the protein levels of SUL in *ago1-27* are similar to those of WT. These results reveal that like animal siRNAs, plant siRNAs mediate translational inhibition.

In addition, siRNA-mediated translational inhibition is also present in *Chlamydomonas reinhardtii*. When an IR transgene was used to silence MAA7, a trptophan β-synthase subunit, ~20% transformants, in which MAA7 protein levels are reduced, displayed no significant change of MAA7 mRNA levels (Cerutti, personal communication). Furthermore, disruption of an exportin 5-like protein in these transformants reduces the accumulation of IR siRNAs, resulting in an increased MAA7 protein levels and unchanged MAA7 mRNA levels (Cerutti, personal communication). These data suggested that IR siRNAs can inhibit translation and raises a question why some siRNAs mediate translation inhibition, while others guide target cleavage (Cerutti, personal communication).

3.4.3 Mechanistic Similarity of Translational Inhibition by miRNAs Between Plants and Animals

Unlike in animals, our understanding on the mechanisms governing translational inhibition by plant miRNAs only begins to emerge. Among ten AGO proteins from Arabidopsis, AGO1 and AGO10 were proposed to be involved in translational inhibition (Brodersen et al. 2008). In a hypomorphic *ago1-27* mutant, the protein levels of several target increase dramatically while the mRNA levels increase only moderately, suggesting that AGO1 may contribute to translational inhibition (Brodersen et al. 2008). AGO10 protein is the closed paralog of AGO1 and might function redundantly with AGO1 in some aspects, because the *ago10* mutant has some similar developmental defects with *ago1*, and *ago1 ago10* double mutations cause lethality. In fact, in a frameshift *ago10* mutant, the protein levels of CSD2, a miR398 target, increased disproportionately higher than the mRNA levels, indicating that AGO10 might be a player of translational inhibition (Brodersen et al. 2008).

The deficiency of miRNA-mediated translational inhibition in *mad5* is caused by a mutation in the *KATANIN1 (KTN1)* gene (Brodersen et al. 2008), which encodes the catalytic subunit of 60 kDa (P60) of the microtubule-severing protein (Burk et al. 2001). KTN1 is an ATPase associated with various cellular activities. *KTN1* severs microtubules in the presence of ATP in vitro (Stoppin-Mellet et al. 2002). Over-expressing *KTN1* in the plant interphase cells releases cortical microtubules and generates motile microtubules that incorporate into bundles, indicating the role of *KTN1* in microtubule dynamics (Stoppin-Mellet et al. 2006). The reduction of protein but not mRNA levels of miRNA targets in the *mad5* and other mutant alleles of KTN1 suggested a role of microtubule dynamics in the miRNA-mediated translational inhibition but not miRNA-mediated cleavage (Brodersen et al. 2008). The involvement of microtubules in miRNA-mediated gene regulation has been documented in animals. For instance, in *C. elegans*, the reduction tubulins by RNAi disrupted miRNA-guided translational inhibition (Parry et al. 2007); in Drosophila, Armitage, a microtubule-associated protein, functions in RISC assembly (Cook et al. 2004). The finding that microtubules are required for miRNA-mediated translational inhibition suggested that plant miRNAs might employ similar mechanisms to direct translational inhibition (Brodersen et al. 2008). In fact, *VSC*, a component of the decapping complex, is required for plant miRNA-mediated translational inhibition, which resembles the function of the components of animal decapping complex in animals such as DCP1, DCP2, and Ge-1 (Brodersen et al. 2008; Eulalio et al. 2007).

3.5   Roles of miRNA-Mediated Translational Inhibition in Plants

3.5.1   *miR172-Mediated Translational Inhibition*

The first plant miRNA that was observed to act through translational inhibition is miR172, which presents in both eudicotyledons and monocotyledons including Arabidopsis, Rice, and others (Aukerman and Sakai 2003; Chen 2004). miR172 represses the expression of a class of *APETALA2 (AP2)-like* transcription factors (Table 3.1), including AP2, TOE1-3, SMZ, and SNZ, which are involved in controlling multiple biological processes in plants such as development, hormone responses, and disease resistance (Park et al. 2002; Schmid et al. 2005; Schwab et al. 2005). In plants, miR172-mediated translational inhibition has been shown to regulate floral patterning, floral determinacy, and flowering time.

3.5.1.1   Roles of miR172-Mediated Translational Inhibition in Floral Patterning

Most angiosperm flowers possess four types of organs, from outside to inside, sepals, petals, stamens, and carpels. They are arranged in a series of whorls. Among them, stamen and carpel are reproductive organs, and sepals and petals are perianth organs. It was proposed that the floral organ identity is specified through the combinatorial activities of three classes of genes, called *A*, *B*, and *C* genes (Jack 2004). Most of these *A*, *B*, and *C* genes are transcription factors. The *A* genes alone specify sepal, the *A* and *B* genes specify petals, the *B* and genes specify stamen, and the *C* genes alone specify carpel. *A* and *C* genes reciprocally repress each other to restrict their domains of activities (Jack 2004). Recent studies have revealed the important role of translational inhibition by miR172 in specifying organ identity, adding a new layer of regulating network of floral organ identity.

APETALA2 (AP2) functions as an *A* gene because loss of function of *ap2* mutation results in the expression of *AGAMOUS (AG)*, a *C* gene, in the outer two whorls, which causes the replacement of perianth organs by reproductive organs (Jack 2004). AP2 protein contains a miR172-binding site and appears to be regulated by miR172 through translational inhibition. Overexpression of miR172 under the control of a cauliflower mosaic virus (CaMV) 35S promoter reduced the AP2 protein levels without significantly changing the accumulation of AP2 mRNAs, suggesting translational regulation of AP2 by miR172 (Aukerman and Sakai 2003; Chen 2004). Consistent with this, the expression of an *AP2* cDNA (AP2m3), in which the binding site of miR172 was abolished, but not wild-type *AP2* cDNA increased AP2 protein levels without affecting *AP2* mRNA levels (Chen 2004). Overexpression of miR172 converted the perianth organs to reproductive organs, which resembles the phenotypes of *ap2* mutant, indicating loss-of-control of AG by AP2 in the outer two whorls (Aukerman and Sakai 2003; Chen 2004). In AP2m3, the expression of AG is reduced in the inner two whorls, and the reproductive organs are replaced with perianth organs (Chen 2004). These results demonstrated the important role of miR172 in controlling floral patterning through translational inhibition of AP2. Unlike in Arabidopsis, overexpression of miR172 from Arabidopsis in *Nicotiana benthaninmana* converts sepal to petal and/or more sepals and petals (Mlotshwa et al. 2006). Although it is not clear how miR172 regulates floral homeotic transformation in *N. benthanimana*, this result indicated that miR172-mediated gene regulation might have different roles in floral patterning in different plant species (Mlotshwa et al. 2006).

In Arabidopsis, the stamen identity determination requires the function of class *B* genes, AP1 and PI, which are present in whirl 2 and 3 (Zhao et al. 2007a). The presence of stamens flanking floral meristem in AP2m3 suggested that the expression domain of AP3 and PI was expanded (Zhao et al. 2007a). In fact, AP3 and PI RNAs are present in all internal stamen primordia in AP2m3 (Zhao et al. 2007a). In addition, PI RNAs also exist in the center of the meristem throughout flower development since its initiation in AP2m3 (Zhao et al. 2007a). These data suggested a role of miR172 in defining the boundary of *B* gene expression domain in whorl 2 and whorl 3 (Zhao et al. 2007a).

The floral meristem terminates after the production of carpel. However, the flowers of AP2m3 plants contain numerous stamens flanking an indeterminate floral meristem, suggesting that miR172 mediated-repression of AP2 is crucial in specifying floral stem cell fate (Zhao et al. 2007a). In Arabidopsis, *WUSCHEL (WUS)*, a homeodomain transcription factor, functions in regulating floral stem cells (Clark 2001). The *WUS* gene is expressed in a small number of cells underneath the stem cell and identifies the overlying cells as stem cells (Clark 2001). Unlike in the wild-type plants, the expression domain of *WUS* in AP2m3 is expanded into the entire meristem and young organ primordia in the very late stage flowers (Zhao et al. 2007a). Introducing a *wus* mutation into AP2m3 completely abolishes the indeterminate phenotypes (Zhao et al. 2007a). These results confirmed the role of miR172 in regulating floral stem cells through WUS pathway (Zhao et al. 2007a).

3.5.1.2 Roles of miR172-Mediated Translational Inhibition in Floral Timing

miR172-mediated translational inhibition also regulates flowering time. *TOE1* and *TOE2* are another two *AP2*-like transcription factors regulated by miR172 (Aukerman and Sakai 2003). Lack of *TOE1* but not *TOE2* causes slightly early flowering in Arabidopsis (Aukerman and Sakai 2003). However, loss-of-function of both *TOE1* and *TOE2* results in a much earlier flowering phenotype, suggesting that *TOE1* and *TOE2* function redundantly as floral timing repressors (Aukerman and Sakai 2003). Overexpression of miR172 reduces the *TOE1* protein levels without significantly affecting its mRNA level, although the cleavage products of TOE1 can be detected

(Aukerman and Sakai 2003). Unlike *TOE1*, the *TOE2* mRNA levels are reduced by overexpression of miR172, indicating a role of miR172-mediated target cleavage (Aukerman and Sakai 2003). The miR172-overexpressing plants display an extremely early flowering phenotype (Aukerman and Sakai 2003). In addition, overexpression of miR172 suppresses the later flowering phenotype caused by overexpression of *TOE1* (Aukerman and Sakai 2003). These data suggested a role of mi172 in controlling flowering time through repressing TOE1 and TOE2 (Aukerman and Sakai 2003). In addition, the *toe1toe2* does not flower as early as *miR172*-overexpressing plants, indicating that other floral repressors are also targets of *miR172* (Aukerman and Sakai 2003).

3.5.2 *miR156-Mediated Translational Inhibition*

miR156 is another conserved plant miRNA family, which has been shown to function as translational inhibitor. miR156 targets a class of plant specific *SQUAMOSA PROMOTER BINDING PROTEIN-LIKE* (*SPL*) transcription factors (Table 3.1), which contain a highly conserved DNA-binding domain (Cardon et al. 1999; Rhoades et al. 2002). Out of 17 SPL genes, 11 are predicted to be targets of miR156 (Rhoades et al. 2002). SPL3, containing a miR156 target site at 3′ UTR, is a member of SPL gene family and functions in floral induction as overexpression of SPL3 cDNA lead to an early flowering phenotype (Cardon et al. 1997). Some studies have suggested that miR156/157 regulates the expression of SPL3 through target-cleavage, because overexpression of miR156 reduces the levels of SPL3 mRNA and the transcripts of SPL3 are increased in *dcl1-12* and *hasty*, which are known mutants deficient in miRNA biogenesis (Park et al. 2005; Schwab et al. 2005). In addition, the cleavage products of SPL3 by miR156 have been detected in Arabidopsis (Schwab et al. 2005). However, recent evidence

Table 3.1 Validated microRNAs inhibiting translation of targets

MicroRNA	Targets	Targets regulated through translational inhibition	Function
miR172	AP2-like transcription factors	AP2, TOE1	Floral patterning, floral meristem, and floral timing
miR156/157	SPL transcription factors	SPL3	Floral timing
miR398	COX5b.1,CSD1, CSD2	CSD1, CSD2	Oxidative stress
miR171	SCL transcription factors	SCL6-IV	Unknown
miR834	CIP4	CIP4	Positive regulator of photomorphogenesis

AP2 APETALA2; *SPL* SQUAMOSA PROMOTER BINDING PROTEIN-LIKE; *COX5b.1* Cytochrome Oxidase Subunit 5b.1; *CSD* COPPER SUPEROXIDE DIMUTASE; *SCL* SCARECROW-like; *CIP4 COP1-interactive partner 4*

suggested that miR156 also represses the expression of SPL3 at translational level (Gandikota et al. 2007). Overexpression of a SPL3 transgene resistant to miR156 or a wild-type SPL3 transgene under the control of 35S promoter in Arabidopsis produced considerable levels of SPL3 transcripts, but only the transcripts resistant to miR156 generated detectable protein levels of SPL3 (Gandikota et al. 2007). In addition, in *mad5*, a miRNA-mediated translational inhibition deficient mutant, the SPL3 protein levels are increased dramatically, while the SPL3 mRNA levels are slightly decreased, suggesting that miR156 negatively regulate SPL3 translation (Brodersen et al. 2008). These results suggested that miR156 is able to regulate the expression of SPL3 through both target-cleavage and translational inhibition.

Overexpression of miR156 caused a later flowering time indicating that miR156 acts as a floral time repressor (Gandikota et al. 2007). Overexpression of a SPL3 transgene resistant to miR156, but not a wild-type SPL3 transgene, induces early flowering time, suggesting that miR156 regulates flowering time through repressing the expression of SPL3 (Gandikota et al. 2007).

3.5.3 *miR398-Mediated Translational Inhibition*

miR398 represents a class of conserved angiosperm miRNAs and has been shown to function through translational inhibition (Table 3.1). In Arabidopsis, miR398 recognizes two *COPPER SUPEROXIDE DISMUTASE* (*CSD*) genes, *CSD1* and *CSD2*, and a *CYTOCHROME C OXIDASE 5b.1* (*COX5b.1*) gene (Bonnet et al. 2004; Jones-Rhoades and Bartel 2004; Sunkar and Zhu 2004). Due to the lack of an antibody recognizing COX5b.1, it is currently unknown whether miR398 regulate COX5b.1 through translational inhibition. Similar to miR156, miR398 regulates *CSD1* and *CSD2* through both translational inhibition and target cleavage (Dugas and Bartel 2008; Sunkar et al. 2006; Yamasaki et al. 2007). Overexpression of miR398 results in a dramatic reduction of proteins levels of *CSD1* and *CSD2* and a moderate reduction of mRNA levels of *CSD1* and *CSD2*, indicating miR398 regulates *CSD1* and *CSD2* through translational inhibition (Dugas and Bartel 2008). Consistent with this result, the *mad* mutants that are deficient in miRNA-mediated translational inhibition display increased protein levels of CDS1 and CDS2 without significantly changing mRNA levels of *CDS1* and *CDS2* (Brodersen et al. 2008). Overexpression of miR398 reduces the mRNA levels of CSD1 and CSD2 indicating that miR398 also functions through target-cleavage (Dugas and Bartel 2008; Sunkar et al. 2006; Yamasaki et al. 2007). Agreeing with this, the mRNA levels of CSD1 and CSD2 increased dramatically in *dcl1-12*, in which the levels of mR398 are reduced (Brodersen et al. 2008). Intriguingly, introducing CDS1 and CSD2 transgenes with altered miR398-binding site increased the accumulation of mRNAs, but not proteins of CDS1 and CDS2, indicating that altering target site complementarity can change miRNA-mediate cleavage to translational inhibition (Dugas and Bartel 2008).

The function of CSD1 and CSD2 is to protect plants from oxidative stress through neutralizing superoxide radicals by releasing molecular oxygen and hydrogen peroxide. It was suggested that CSD2 might protect plants from oxidative stress generated by photosynthetic activities (Kliebenstein et al. 1998). Their function requires copper as an essential cofactor (Kliebenstein et al. 1998). Several studies have suggested that miR398 regulates CSD1 and CSD2 in response to oxidative stress or copper availability (Sunkar et al. 2006; Dugas and Bartel 2008; Yamasaki et al. 2007). It has been observed that the levels of miR398 are positively regulated by sucrose (Dugas and Bartel 2008). Sugar is able to inhibit photosynthesis, which produces reactive-oxygen species (ROS) (Dugas and Bartel 2008). Therefore, the induction of miR398 by sucrose might reflect that plants reduce the levels of CSD1and CSD2 through enforcing miR398-mediated regulation in response to reduced oxidative stress (Dugas and Bartel 2008). In addition, the miR398 levels are reduced by copper supplement and are increased by copper limitation suggesting copper regulates the levels of CSD1 and CSD2 through releasing or enforcing miR398-mediated regulation (Dugas and Bartel 2008; Sunkar et al. 2006; Yamasaki et al. 2007).

3.5.4 Other miRNA-Mediated Translational Inhibition

Beyond miR172, miR156, and miR398, some other miRNAs, including miR171 and miR834 are also involved in translational inhibition. miR171 targets several members of the SCARECROW-like (SCL) family of putative transcription factors, which are involved in numerous development processes (Brodersen et al. 2008). It has been shown that miR171 regulates SCL6-IV through target cleavage (Llave et al. 2002b). However, disruption of translational control in *mad* mutants lead to increased protein levels without significant effect on the mRNA levels (Brodersen et al. 2008). miR834 is identified only in Arabidopsis and targets *COP1-interactive partner 4* (CIP4), which is a putative transcription factor (Fahlgren et al. 2007). Both *mad* mutants and *dcl1-12* mutant displayed increased CIP4 protein levels and WT mRNA levels, suggesting miR834 regulates CIP4 through translational inhibition (Brodersen et al. 2008).

3.6 Conclusion

Although it has been established that translational inhibition is a common acting mechanism of plant miRNAs, many aspects of translational inhibition remain unclear. Unlike in animals, the mechanism governing miRNA-mediated translational inhibition is little known. Given the fact that both plant and animal miRNA-mediated translational inhibition requires microtubule network and P-body components, plants and animals might have some mechanic similarities of miRNA mediated translational inhibition (Brodersen et al. 2008). It was proposed that the complementarity between miRNAs and targets determines the miRNA activity. However, the same plant miRNA can act through both target cleavage and translational inhibition, raising the question how cells make choices between these two activities (Brodersen et al. 2008). It remains unclear whether these two activities coexist or are temporarily and/or spatially separated (Brodersen et al. 2008). AGO1 performs both cleavage activity and translational inhibition raising a question how another activity is repressed when one activity is ongoing (Brodersen et al. 2008).

Detailed studies on three miRNAs have shown that miRNA mediated translational inhibition are involved in controlling floral patterning, floral timing, and stress responses. Given hundreds of miRNAs have been identified in plants, testing the functional roles of translational inhibition by these miRNAs represents another challenge.

Acknowledgments We thank Dr. Xuemei Chen from University of California, Riverside, for critical comments on this manuscript and Dr. Heriberto D. Cerutti from University of Nebraska, Lincoln, for sharing unpublished data. We also apologize to colleagues whose works were not discussed here due to space limitation.

References

Adenot X, Elmayan T, Lauressergues D, Boutet S, Bouche N, Gasciolli V, Vaucheret H (2006) DRB4-dependent TAS3 trans-acting siRNAs control leaf morphology through AGO7. Curr Biol 16:927–932

Allen E, Xie Z, Gustafson AM, Carrington JC (2005) microRNA-directed phasing during trans-acting siRNA biogenesis in plants. Cell 121:207–221

Aukerman MJ, Sakai H (2003) Regulation of flowering time and floral organ identity by a MicroRNA and its APETALA2-like target genes. Plant Cell 15:2730–2741

Bao N, Lye KW, Barton MK (2004) MicroRNA binding sites in Arabidopsis class III HD-ZIP mRNAs are required for methylation of the template chromosome. Dev Cell 7:653–662

Bartel DP (2004) MicroRNAs: genomics, biogenesis, mechanism, and function. Cell 116:281–297

Baumberger N, Baulcombe DC (2005) Arabidopsis ARGONAUTE1 is an RNA Slicer that selectively recruits microRNAs and short interfering RNAs. Proc Natl Acad Sci U S A 102:11928–11933

Bohmert K, Camus I, Bellini C, Bouchez D, Caboche M, Benning C (1998) AGO1 defines a novel locus of Arabidopsis controlling leaf development. EMBO J 17:170–180

Bonnet E, Wuyts J, Rouze P, Van de Peer Y (2004) Detection of 91 potential conserved plant microRNAs in Arabidopsis thaliana and Oryza sativa identifies important target genes. Proc Natl Acad Sci U S A 101:11511–11516

Borsani O, Zhu J, Verslues PE, Sunkar R, Zhu JK (2005) Endogenous siRNAs derived from a pair of natural cis-antisense transcripts regulate salt tolerance in Arabidopsis. Cell 123:1279–1291

Brodersen P, Sakvarelidze-Achard L, Bruun-Rasmussen M, Dunoyer P, Yamamoto YY, Sieburth L, Voinnet O (2008) Widespread translational inhibition by plant miRNAs and siRNAs. Science 320:1185–1190

Burk D H, Liu B, Zhong R, Morrison W H and Ye ZH (2001) A katanin-like protein regulates normal cell wall biosynthesis and cell elongation. Plant Cell 13:807-827

Cardon G, Hohmann S, Nettesheim K, Saedler H, Huijser P (1997) Functional analysis of the Arabidopsis thaliana SBP-box gene SPL3: a novel gene involved in the floral transition. Plant J 12:367–377

Cardon G, Hohmann S, Klein J, Nettesheim K, Saedler H, Huijser P (1999) Molecular characterisation of the Arabidopsis SBP-box genes. Gene 237:91–104

Chen X (2004) A microRNA as a translational repressor of APETALA2 in Arabidopsis flower development. Science 303:2022–2025

Chen X (2005) MicroRNA biogenesis and function in plants. FEBS Lett 579:5923–5931

Chen X (2008) A silencing safeguard: links between RNA silencing and mRNA processing in Arabidopsis. Dev Cell 14:811–812

Clark SE (2001) Cell signalling at the shoot meristem. Nat Rev Mol Cell Biol 2:276–284

Cook HA, Koppetsch BS, Wu J, Theurkauf WE (2004) The Drosophila SDE3 homolog armitage is required for oskar mRNA silencing and embryonic axis specification. Cell 116:817–829

Dong Z, Han MH, Fedoroff N (2008) The RNA-binding proteins HYL1 and SE promote accurate in vitro processing of pri-miRNA by DCL1. Proc Natl Acad Sci U S A 105:9970–9975

Dugas DV, Bartel B (2008) Sucrose induction of Arabidopsis miR398 represses two Cu/Zn superoxide dismutases. Plant Mol Biol 67:403–417

Eulalio A, Rehwinkel J, Stricker M, Huntzinger E, Yang SF, Doerks T, Dorner S, Bork P, Boutros M, Izaurralde E (2007) Target-specific requirements for enhancers of decapping in miRNA-mediated gene silencing. Genes Dev 21:2558–2570

Fahlgren N, Howell MD, Kasschau KD, Chapman EJ, Sullivan CM, Cumbie JS, Givan SA, Law TF, Grant SR, Dangl JL, Carrington JC (2007) High-throughput sequencing of Arabidopsis microRNAs: evidence for frequent birth and death of MIRNA genes. PLoS ONE 2:e219

Fang Y, Spector DL (2007) Identification of nuclear dicing bodies containing proteins for microRNA biogenesis in living Arabidopsis plants. Curr Biol 17:818–823

Gandikota M, Birkenbihl RP, Hohmann S, Cardon GH, Saedler H, Huijser P (2007) The miRNA156/157 recognition element in the 3' UTR of the Arabidopsis SBP box gene SPL3 prevents early flowering by translational inhibition in seedlings. Plant J 49:683–693

Gregory BD, O'Malley RC, Lister R, Urich MA, Tonti-Filippini J, Chen H, Millar AH, Ecker JR (2008) A link between RNA metabolism and silencing affecting Arabidopsis development. Dev Cell 14:854–866

Han MH, Goud S, Song L, Fedoroff N (2004) The Arabidopsis double-stranded RNA-binding protein HYL1 plays a role in microRNA-mediated gene regulation. Proc Natl Acad Sci U S A 101:1093–1098

Hernandez-Pinzon I, Yelina NE, Schwach F, Studholme DJ, Baulcombe D, Dalmay T (2007) SDE5, the putative homologue of a human mRNA export factor, is required for transgene silencing and accumulation of trans-acting endogenous siRNA. Plant J 50:140–148

Herr AJ, Jensen MB, Dalmay T, Baulcombe DC (2005) RNA polymerase IV directs silencing of endogenous DNA. Science 308:118–120

Jack T (2004) Molecular and genetic mechanisms of floral control. Plant Cell 16(Suppl):S1–S17

Jones-Rhoades MW, Bartel DP (2004) Computational identification of plant microRNAs and their targets, including a stress-induced miRNA. Mol Cell 14:787–799

Kanno T, Huettel B, Mette MF, Aufsatz W, Jaligot E, Daxinger L, Kreil DP, Matzke M, Matzke AJ (2005) Atypical RNA polymerase subunits required for RNA-directed DNA methylation. Nat Genet 37:761–765

Kasschau KD, Fahlgren N, Chapman EJ, Sullivan CM, Cumbie JS, Givan SA, Carrington JC (2007) Genome-wide profiling and analysis of Arabidopsis siRNAs. PLoS Biol 5:e57

Katiyar-Agarwal S, Morgan R, Dahlbeck D, Borsani O, Villegas A Jr, Zhu JK, Staskawicz BJ, Jin H (2006) A pathogen-inducible endogenous siRNA in plant immunity. Proc Natl Acad Sci U S A 103:18002–18007

Katiyar-Agarwal S, Gao S, Vivian-Smith A, Jin H (2007) A novel class of bacteria-induced small RNAs in Arabidopsis. Genes Dev 21:3123–3134

Kliebenstein DJ, Monde RA, Last RL (1998) Superoxide dismutase in Arabidopsis: an eclectic enzyme family with disparate regulation and protein localization. Plant Physiol 118:637–650

Kurihara Y, Watanabe Y (2004) Arabidopsis micro-RNA biogenesis through Dicer-like 1 protein functions. Proc Natl Acad Sci U S A 101:12753–12758

Laubinger S, Sachsenberg T, Zeller G, Busch W, Lohmann JU, Ratsch G, Weigel D (2008) Dual roles of the nuclear cap-binding complex and SERRATE in pre-mRNA splicing and microRNA processing in Arabidopsis thaliana. Proc Natl Acad Sci U S A 105:8795–8800

Lee RC, Feinbaum RL, Ambros V (1993) The C. elegans heterochronic gene lin-4 encodes small RNAs with antisense complementarity to lin-14. Cell 75:843–854

Li J, Yang Z, Yu B, Liu J, Chen X (2005) Methylation protects miRNAs and siRNAs from a 3'-end uridylation activity in Arabidopsis. Curr Biol 15:1501–1507

Llave C, Kasschau KD, Rector MA, Carrington JC (2002a) Endogenous and silencing-associated small RNAs in plants. Plant Cell 14:1605–1619

Llave C, Xie Z, Kasschau KD, Carrington JC (2002b) Cleavage of Scarecrow-like mRNA targets directed by a class of Arabidopsis miRNA. Science 297:2053–2056

Lobbes D, Rallapalli G, Schmidt DD, Martin C, Clarke J (2006) SERRATE: a new player on the plant microRNA scene. EMBO Rep 7:1052–1058

Lu C, Kulkarni K, Souret FF, MuthuValliappan R, Tej SS, Poethig RS, Henderson IR, Jacobsen SE, Wang W, Green PJ, Meyers BC (2006) MicroRNAs and other small RNAs enriched in the Arabidopsis RNA-dependent RNA polymerase-2 mutant. Genome Res 16:1276–1288

Mallory AC, Vaucheret H (2006) Functions of microRNAs and related small RNAs in plants. Nat Genet 38(Suppl):S31–S36

Mette MF, van der Winden J, Matzke M, Matzke AJ (2002) Short RNAs can identify new candidate transposable element families in Arabidopsis. Plant Physiol 130:6–9

Meyers BC, Souret FF, Lu C, Green PJ (2006) Sweating the small stuff: microRNA discovery in plants. Curr Opin Biotechnol 17:139–146

Mi S, Cai T, Hu Y, Chen Y, Hodges E, Ni F, Wu L, Li S, Zhou H, Long C, Chen S, Hannon GJ, Qi Y (2008) Sorting of small RNAs into Arabidopsis argonaute complexes is directed by the 5' terminal nucleotide. Cell 133:116–127

Mlotshwa S, Yang Z, Kim Y, Chen X (2006) Floral patterning defects induced by Arabidopsis APETALA2 and microRNA172 expression in Nicotiana benthamiana. Plant Mol Biol 61: 781–793

Montgomery TA, Howell MD, Cuperus JT, Li D, Hansen JE, Alexander AL, Chapman EJ, Fahlgren N, Allen E, Carrington JC (2008) Specificity of ARGONAUTE7-miR390 interaction and dual functionality in TAS3 trans-acting siRNA formation. Cell 133:128–141

Morel JB, Godon C, Mourrain P, Beclin C, Boutet S, Feuerbach F, Proux F, Vaucheret H (2002) Fertile hypomorphic ARGONAUTE (ago1) mutants impaired in post-transcriptional gene silencing and virus resistance. Plant Cell 14:629–639

Nonomura K, Morohoshi A, Nakano M, Eiguchi M, Miyao A, Hirochika H, Kurata N (2007) A germ cell specific gene of the ARGONAUTE family is essential for the progression of premeiotic mitosis and meiosis during sporogenesis in rice. Plant Cell 19:2583–2594

Onodera Y, Haag JR, Ream T, Nunes PC, Pontes O, Pikaard CS (2005) Plant nuclear RNA polymerase IV mediates siRNA and DNA methylation-dependent heterochromatin formation. Cell 120:613–622

Park W, Li J, Song R, Messing J, Chen X (2002) CARPEL FACTORY, a Dicer homolog, and HEN1, a novel protein, act in microRNA metabolism in Arabidopsis thaliana. Curr Biol 12: 1484–1495

Park MY, Wu G, Gonzalez-Sulser A, Vaucheret H, Poethig RS (2005) Nuclear processing and export of microRNAs in Arabidopsis. Proc Natl Acad Sci U S A 102:3691–3696

Parry DH, Xu J, Ruvkun G (2007) A whole-genome RNAi Screen for C. elegans miRNA pathway genes. Curr Biol 17:2013–2022

Pontier D, Yahubyan G, Vega D, Bulski A, Saez-Vasquez J, Hakimi MA, Lerbs-Mache S, Colot V, Lagrange T (2005) Reinforcement of silencing at transposons and highly repeated sequences requires the concerted action of two distinct RNA polymerases IV in Arabidopsis. Genes Dev 19:2030–2040

Qi Y, He X, Wang XJ, Kohany O, Jurka J, Hannon GJ (2006) Distinct catalytic and non-catalytic roles of ARGONAUTE4 in RNA-directed DNA methylation. Nature 443:1008–1012

Rajagopalan R, Vaucheret H, Trejo J, Bartel DP (2006) A diverse and evolutionarily fluid set of microRNAs in Arabidopsis thaliana. Genes Dev 20:3407–3425

Ramachandran V, Chen X (2008a) Degradation of microRNAs by a family of exoribonucleases in Arabidopsis. Science 321:1490–1492

Ramachandran V, Chen X (2008b) Small RNA metabolism in Arabidopsis. Trends Plant Sci 13:368–374

Reinhart BJ, Weinstein EG, Rhoades MW, Bartel B, Bartel DP (2002) MicroRNAs in plants. Genes Dev 16:1616–1626

Rhoades MW, Reinhart BJ, Lim LP, Burge CB, Bartel B, Bartel DP (2002) Prediction of plant microRNA targets. Cell 110:513–520

Schmid M, Davison TS, Henz SR, Pape UJ, Demar M, Vingron M, Scholkopf B, Weigel D, Lohmann JU (2005) A gene expression map of Arabidopsis thaliana development. Nat Genet 37:501–506

Schwab R, Palatnik JF, Riester M, Schommer C, Schmid M, Weigel D (2005) Specific effects of microRNAs on the plant transcriptome. Dev Cell 8:517–527

Song L, Han MH, Lesicka J, Fedoroff N (2007) Arabidopsis primary microRNA processing proteins HYL1 and DCL1 define a nuclear body distinct from the Cajal body. Proc Natl Acad Sci U S A 104:5437–5442

Stoppin-Mellet V, Gaillard J, Vantard M (2002) Functional evidence for in vitro microtubule severing by the plant katanin homologue. Biochem J 365:337–342

Stoppin-Mellet V, Gaillard J, Vantard M (2006) Katanin's severing activity favors bundling of cortical microtubules in plants. Plant J 46:1009–1017

Sunkar R, Zhu JK (2004) Novel and stress-regulated microRNAs and other small RNAs from Arabidopsis. Plant Cell 16:2001–2019

Sunkar R, Kapoor A, Zhu JK (2006) Posttranscriptional induction of two Cu/Zn superoxide dismutase genes in Arabidopsis is mediated by downregulation of miR398 and important for oxidative stress tolerance. Plant Cell 18:2051–2065

Takeda A, Iwasaki S, Watanabe T, Utsumi M, Watanabe Y (2008) The mechanism selecting the guide strand from small RNA duplexes is different among argonaute proteins. Plant Cell Physiol 49:493–500

Vaucheret H (2008) Plant ARGONAUTES. Trends Plant Sci 13:350–358

Vazquez F, Vaucheret H, Rajagopalan R, Lepers C, Gasciolli V, Mallory AC, Hilbert JL, Bartel DP, Crete P (2004) Endogenous trans-acting siRNAs regulate the accumulation of Arabidopsis mRNAs. Mol Cell 16:69–79

Xie Z, Johansen LK, Gustafson AM, Kasschau KD, Lellis AD, Zilberman D, Jacobsen SE, Carrington JC (2004) Genetic and functional diversification of small RNA pathways in plants. PLoS Biol 2:E104

Yamasaki H, Abdel-Ghany SE, Cohu CM, Kobayashi Y, Shikanai T, Pilon M (2007) Regulation of copper homeostasis by micro-RNA in Arabidopsis. J Biol Chem 282:16369–16378

Yang L, Liu Z, Lu F, Dong A, Huang H (2006a) SERRATE is a novel nuclear regulator in primary microRNA processing in Arabidopsis. Plant J 47:841–850

Yang Z, Ebright YW, Yu B, Chen X (2006b) HEN1 recognizes 21–24 nt small RNA duplexes and deposits a methyl group onto the 2' OH of the 3' terminal nucleotide. Nucl Acids Res 34: 667–675

Yoshikawa M, Peragine A, Park MY, Poethig RS (2005) A pathway for the biogenesis of trans-acting siRNAs in Arabidopsis. Genes Dev 19:2164–2175

Yu B, Yang Z, Li J, Minakhina S, Yang M, Padgett RW, Steward R, Chen X (2005) Methylation as a crucial step in plant microRNA biogenesis. Science 307:932–935

Yu B, Bi L, Zheng B, Ji L, Chevalier D, Agarwal M, Ramachandran V, Li W, Lagrange T, Walker JC, Chen X (2008) The FHA domain proteins DAWDLE in Arabidopsis and SNIP1 in humans act in small RNA biogenesis. Proc Natl Acad Sci U S A 105:10073–10078

Zhang X, Henderson IR, Lu C, Green PJ, Jacobsen SE (2007) Role of RNA polymerase IV in plant small RNA metabolism. Proc Natl Acad Sci U S A 104:4536–4541

Zhao L, Kim Y, Dinh TT, Chen X (2007a) miR172 regulates stem cell fate and defines the inner boundary of APETALA3 and PISTILLATA expression domain in Arabidopsis floral meristems. Plant J 51:840–849

Zhao T, Li G, Mi S, Li S, Hannon GJ, Wang XJ, Qi Y (2007b) A complex system of small RNAs in the unicellular green alga Chlamydomonas reinhardtii. Genes Dev 21:1190–1203

Zheng X, Zhu J, Kapoor A, Zhu JK (2007) Role of Arabidopsis AGO6 in siRNA accumulation, DNA methylation and transcriptional gene silencing. EMBO J 26:1691–1701

Chapter 4
Regulation of p27^{kip1} mRNA Expression by MicroRNAs

Aida Martínez-Sánchez and Fátima Gebauer

Contents

Abstract p27^{kip1} (p27) is a cell cycle inhibitor and tumor suppressor whose expression is highly regulated in the cell. Low levels of p27 have been associated with poor prognosis in cancer. Recently, several microRNAs have been described to control p27 expression in various tumor types. In this chapter, we will provide an overview on the role of microRNAs in cancer, and will discuss how microRNAs regulate p27 expression and the implications for tumor progression.

4.1 Introduction

MicroRNAs (miRNAs) are conserved noncoding RNA molecules of about 22 nucleotides that regulate gene expression post-transcriptionally. Generally, miRNAs recognize complementary sequences present in multiple copies in the 3′ untranslated region (UTR) of target mRNAs and repress their translation (reviewed in Filipowicz et al. 2008). Translational repression is frequently accompanied by destabilization of the mRNA target, and, in some cases, mRNA degradation plays an important role in regulation (see Chap. 2 in this volume). Despite considerable efforts, the molecular

A. Martínez-Sánchez and F. Gebauer (✉)
Centre de Regulació Genòmica (CRG-UPF), Gene Regulation Programme,
Dr Aiguader 88, 08003 Barcelona, Spain
e-mail: fatima.gebauer@crg.es

R.E. Rhoads (ed.), *miRNA Regulation of the Translational Machinery*,
Progress in Molecular and Subcellular Biology 50,
DOI 10.1007/978-3-642-03103-8_4, © Springer-Verlag Berlin Heidelberg 2010

mechanisms of translational repression by miRNAs are unclear at present. This aspect of miRNA regulation is discussed in detail elsewhere in this volume (see Chaps. 1 and 7 in this volume, chapter by Nilsen) as well as in a number of recent reviews (Standart and Jackson 2007; Jackson and Standart 2007; Eulalio et al. 2008; Filipowicz et al. 2008; Richter 2008), and will not be discussed further here.

Since their discovery, miRNAs have been shown to play a role in a wide variety of biological processes, including embryonic development, morphogenesis, proliferation, differentiation, inflammation, and apoptosis (reviewed in Bushati and Cohen 2007; Bueno et al. 2008). During these processes, miRNAs can act as rheostats that fine-tune the expression of many mRNAs in the cell or as robust regulators of specific target genes. One of these genes is the tumor suppressor p27^{kip1} (p27), a cell cycle inhibitor whose deregulation contributes to tumor progression. In this chapter, we will focus on the role of miRNAs in cancer, and particularly on the expression of p27.

4.2 miRNAs in Cancer

Cancer is a complex genetic disease caused by accumulation of mutations leading to uncontrolled cell growth and proliferation. Classically, the causes of tumorigenesis have been attributed to alteration of protein-coding genes, but recent evidences indicate that changes in miRNA expression also contribute to tumor formation. For example, global depletion of miRNAs by impairing miRNA processing increases cellular transformation, and ectopic expression of miRNAs such as miR-155 or the miR-17-92 cluster accelerates tumor development (Kumar et al. 2007; He et al. 2005; Costinean et al. 2006). miRNAs can be used to distinguish normal from tumor tissue, cancer type, stage, and other clinical variables. In fact, profiling experiments have shown that miRNA changes are better predictors of tumor type than mRNA changes and have led to the identification of miRNA signatures for specific types of cancer (reviewed in Lee and Dutta 2009). miRNA expression profiles are not only a useful diagnosis but also a prognosis tool, as correlations have been established between the expression of certain miRNAs and the survival of patients (e.g., Yanaihara et al. 2006; Takamizawa et al. 2004). Multiple mechanisms underlie the widespread disruption of miRNA expression in tumors, including: (1) the alteration of the genomic region where miRNA genes are located (Calin et al. 2004); (2) the epigenetic modification – DNA methylation or histone deacetylation – of miRNA loci (Lehmann et al. 2008); (3) the aberrant transcription of miRNA precursors (He et al. 2007); and (4) the abnormal expression of factors involved in miRNA processing (Karube et al. 2005). In addition, the expression of modulators of miRNA function could also play a role (Kedde et al. 2007).

In order to understand the multiple functions of miRNAs in biological processes, it becomes essential to identify their targets. In silico predictions of miRNA targets are often inaccurate because the miRNA/mRNA interaction basically relies on a limited sequence length: the eight nucleotides of the miRNA seed region. In addition, mRNA recognition is influenced by the sequence context around the target site and by factors that may block miRNA binding. Furthermore, although definitely useful, screenings based on over-expression of miRNAs usually yield a large number of false positives due to off-target effects. Alternative approaches consisting of immunoprecipitation of miRNP-associated transcripts or mRNA profiling after miRNA depletion are likely to improve the number of bona fide targets. To date, much information has accumulated about putative miRNA/mRNA pairs, but only a few have been experimentally validated. Validation is considered here in a rigorous sense, and includes reporter assays showing that the miRNA directly represses mRNA expression and that mutation of the miRNA binding sites abrogates regulation. Table 4.1 summarizes those validated pairs with a function in tumor development.

Table 4.1 Validated miRNA/ mRNA pairs in tumor progression

miRNA	mRNA target[a]	Mechanism[b]	Tumor	References
let-7	HMGA2	R	Ovary, lung	Shell et al. (2007)
				Lee and Dutta (2009)
				Mayr et al. (2007)
	NF2	nd	Cholangio carcinoma	Meng et al. (2007)
let-7g	k-Ras, c-Myc	nd	Various	Kumar et al. (2007)
miR-9,-125a, -125b	trkC	nd	Neuroblastoma	Laneve et al. (2007)
miR-10b	HOXD 10	T	Breast	Ma et al. (2007)
miR-16-1, -15a	Bcl-2	T	Leukemia	Cimmino et al. (2005)
miR-17-5p	AIB1	T	Breast	Hossain et al. (2006)
	p21	R	Neuroblastoma	Fontana et al. (2008)
miR-17-5p, -20	TβRII	R		Tagawa et al. (2007)
miR-20a	E2F1, 2, 3	T		Sylvestre et al. (2007)
miR-21	Pdcd4	T	Colon, breast	Asangani et al. (2007)
				Lu et al. (2008)
	TMP1	T		Zhu et al. (2007)
miR-29b	Mcl-1	T	Cholangio carcinoma	Mott et al. (2007)
miR-34a	Bcl-2	nd	Lung	Bommer et al. (2007)
miR-124a	CDK6	T	Lung, colon Medulloblastoma	Lujambio et al. (2007) Pierson et al. (2008)
miR-206	ERα	R	Breast	Adams et al. (2007)
miR-221,-222	p27	T	Glioblastoma, prostate	Le Sage et al. (2007) Galardi et al. (2007)
miR-372,-373	LATS2	T, R	Testis	Voorhoeve et al. (2006)
miR-378	Fus-1, Sufu	T	Glioblastoma	Lee et al. (2007)
BART cluster[c]	LMP1[(c)]	T	Nasopharynge	Lo et al. (2007)

[a]*HMGA2* High mobility group A2; *NF2* Neurofibromatosis 2; *trKC* tropomyosin-related kinase C; *AIB1* Amplified in breast cancer 1; *TβRII* Transforming growth factor beta-receptor type 2; *Pdcd4* Programmed cell death 4; *TMP1* Tropomyosin 1; *ERα* Estrogen receptor alpha; *LATS2* Large tumor suppressor homolog 2; *LMP1* Latent membrane protein 1
[b]*R* RNA degradation; *T* translational control; *nd* not determined
[c]Encoded by the EBV (Epstein–Barr) virus genome

miRNAs can either be up- or downregulated in tumors. let-7, one of the founding miRNA members, inhibits the expression of the oncogenes k-Ras, c-Myc, and HMGA2. Consistently, the levels of let-7 are reduced in several tumors, while they increase in differentiated tissues (e.g., Shell et al. 2007). Similarly, miR-15a and miR-16-1 regulate the expression of the mRNA encoding the antiapoptotic protein Bcl-2, and their levels are reduced in chronic lymphocytic leukemia (Cimmino et al. 2005). Conversely, the levels of the miR-17-92 cluster are high in primary neuroblastoma tumors, especially in those with poor prognosis. miR-17-92 inhibits the expression of the cell cycle inhibitor p21, leading to increased cell proliferation (Fontana et al. 2008). miR-10b and miR-21 are also over-expressed in cancer and promote invasion and metastasis via the translational repression of HOXD10, and TPM1 and PDCD4 mRNAs, respectively (Zhu et al. 2007; Lu et al. 2008; Ma et al. 2007). Interestingly, the related miRNA, miR-10a, also over-expressed in tumors, has been proposed to bind to the 5′ UTR of ribosomal protein mRNAs and to increase their translation, a function that may contribute to activate global protein synthesis and growth of transformed cells (Ørom et al. 2008). Thus, miRNAs can behave as either oncogenes or tumor suppressors depending on the cell type.

A recent screening has identified miR-221 and miR-222 as regulators of the expression of the tumor suppressor p27. The broad range of functions in which this protein is involved marks p27 as a prominent mediator of miRNA effects in cancer. In the following sections, we will discuss the roles of p27 and the implications of its regulation by miRNAs in tumorigenesis.

4.3 Role of p27 in Tumor Progression

The cell division cycle is driven by the alternate activity of cyclin-dependent kinases (CDKs). The activity of CDKs, in turn, is regulated by their association with regulatory cyclin subunits, the phosphorylation of their catalytic kinase sub-units, their subcellular localization, and their binding to regulatory proteins called cyclin-dependent kinase inhibitors (CKIs) (reviewed in Malumbres and Barbacid 2007; Sherr and Roberts 2004). p27 is a member of the Cip/Kip family of CKIs; it binds into the catalytic cleft of the cyclin/CDK complex preventing ATP recognition (Russo et al. 1996). Typically, p27 inhibits the G1/S transition by binding to the S-phase promoting kinases, cyclin E/CDK2 and cyclin A/CDK2, although it has also been reported to inhibit the G2/M transition by regulating the activity of CDK1 (Aleem et al. 2005). The antiproliferative role of p27 depends on its localization in the nucleus, coincident with its target kinases. Cytoplasmic p27 performs alternative functions, including the regulation of cytoskeletal structure and cell migration. p27 inhibits the activity of the GTPase RhoA, which is necessary for the adhesion of cells to the substrate, thereby promoting cell motility (Besson et al. 2004). In addition, p27 is necessary for the complete differentiation of several cell types and has been shown to modulate apoptosis (e.g., Baldassarre et al. 1999; Nguyen et al. 2006; Bryja et al. 2004; Philipp-Staheli et al. 2001). The function of p27 in apoptosis appears to be highly dependent on the experimental model, as some studies show a proapoptotic effect of the protein while others report an antiapoptotic role (reviewed in Borriello et al. 2007; Besson et al. 2008). It is interesting to note that the processes controlled by p27 (cell proliferation, differentiation, migration, and apoptosis) are often deregulated in cancer.

According to the important roles of p27 in the cell, expression of p27 is tightly regulated in a cell-type and condition-specific manner. The importance of maintaining adequate levels of p27 is illustrated by the phenotypes of the p27 knock-out and heterozygous mice. Null mice develop increased body size with multiple organ hyperplasia, show greater predisposition to induced tumorigenesis and develop pituitary tumors spontaneously (Fero et al. 1996; Kiyokawa et al. 1996; Nakayama et al. 1996). p27 heterozygous mice, containing half the amount of protein of wild type animals, show an intermediate phenotype (Fero et al. 1998). Molecular analysis of tumors from these mice showed that the remaining wild type allele was neither mutated nor silenced, indicating that p27 is haplo-insufficient for tumor suppression. p27 is also associated with spontaneous tumorigenesis in humans, since many human cancers express decreased amounts of p27 compared to normal tissues. Moreover, low levels of p27 frequently correlate with increased tumor aggressiveness and poor clinical outcome (Chu et al. 2008). Usually, the p27 gene is not mutated or deleted in cancer, but tumors associate with altered post-transcriptional regulation. Although in most cell types p27 plays an antitumorigenic role, elevated p27 levels are not always beneficial. Indeed, according to the function of p27 in cell migration, there is a positive correlation between elevated cytoplasmic p27 levels and invasiveness of a number of tumors, such as melanoma, leukemia, and breast, cervix, and uterus carcinomas (Denicourt et al. 2007; Dellas et al. 1998; Vrhovac et al. 1998; Kouvaraki et al. 2002; Watanabe et al. 2002). In addition, p27 might contribute to the resistance of some tumor cells to chemotherapy-induced apoptosis (Blain et al. 2003). Thus, given the complexity of p27 functions, an accurate knowledge of the mechanisms controlling p27 expression in each cell type is necessary to develop successful therapies against cancer.

4.4 Regulation of p27 Expression: Role of miRNAs

Expression of p27 is regulated at multiple levels, including transcription, mRNA stability, translation, proteolysis, and subcellular localization. Several mechanisms of regulation may coexist in a single cell depending on the cell type, the extracellular stimuli, and the biological circumstances (reviewed in le Sage et al. 2007b; Chu et al. 2008; Borriello et al. 2007; Vervoorts and Lüscher 2008; Koff 2006).

Translational regulation of p27 mRNA has emerged as a prominent mechanism to regulate p27 expression during differentiation, quiescence, and cancer progression. Early reports indicated that the translation of p27 mRNA increased in Hela cells arrested in G1 by treatment with lovastatin, in quiescent, contact-inhibited fibroblasts and in differentiated human promyelocytic leukemia (HL60) cells (Hengst and Reed

1996; Millard et al. 1997). Subsequent studies showed that the 5′ UTR of p27 mRNA contains a number of regulatory features that could allow p27 expression independently on the fate of most cellular mRNAs. For example, an upstream open reading frame was proposed to contribute to translational regulation of p27 during the cell cycle (Göpfert et al. 2003). In addition, several groups have reported the presence of an internal ribosome entry site (IRES) that promotes p27 translation in conditions where cap-dependent translation of most cellular messages is compromised (Miskimins et al. 2001; Kullmann et al. 2002; Cho et al. 2005; Jiang et al. 2007). The IRES is recognized by the proteins PTB, HuR, and hnRNPC1/C2 (Millard et al. 2000; Kullmann et al. 2002; Cho et al. 2005). However, the role of these proteins in p27 mRNA translation is unclear and the existence of the IRES has been recently disputed, as the detected activity was attributed to cryptic promoters in p27 5′ UTR (Liu et al. 2005; Cuesta et al. 2009).

More recently, miRNAs have appeared as chief regulators of p27 mRNA expression. The first indications that miRNAs could play a role in the expression of p27 were obtained in *Drosophila* (Hatfield et al. 2005). Inhibition of miRNA processing by mutation of dicer-1 resulted in delayed G1/S transition of germ-line stem cells, a process controlled by the p27 homolog Dacapo. Reduction of Dacapo levels partially rescued dicer-1 mutants and over-expression of Dacapo resembled dicer-1 mutations, suggesting that an adequate processing of miRNAs was necessary to repress Dacapo. Importantly, expression of a Dacapo transgene lacking a region of the 3′ UTR predicted to contain miRNA binding sites was not affected by dicer-1 mutations (Hatfield et al. 2005). These results suggested that the expression of Dacapo was regulated by miRNAs binding to the 3′ UTR. Indeed, mammalian p27 mRNA translation is regulated via its 3′ UTR in Hela, MDA468, and 3T3 cells (Millard et al. 2000; Vidal et al. 2002; Gonzalez et al. 2003). Significantly, similar to *Drosophila*, depletion of dicer in human glioblastoma cells increases p27 levels and decreases proliferation (Gillies and Lorimer 2007).

An elegant functional screening identified miRNAs that regulate p27 expression (le Sage et al. 2007a). In this screening, Hela cells expressing the GFP coding sequence fused to the 3′ UTR of p27 were transduced with a miRNA library, and cells expressing low levels of GFP were selected. The particular miRNA expressed in these cells was identified as miR-221. This miRNA also repressed the expression of endogenous p27, while p27 mRNA levels and the steady state of p27 protein remained unchanged. These results established that miR-221 represses the translation of p27 mRNA. Bioinformatics analysis predicted two target sites for miR-221 and the related miRNA miR-222 in the 3′ UTR of p27. miR-222 is encoded in the same genomic cluster as miR-221 and contains the same seed sequence. Validations using luciferase reporters showed that over-expressed miR-221/222 repressed the expression of transcripts containing wild type, but not mutated target sites. Conversely, mutation of the miRNA seed sequence in the miRNA- expressing vector abolished repression. In addition, antagomirs (antisense RNA oligos containing a molecule of cholesterol at the 5′ end and 2′-O-methylated at every nucleotide) against miR-221/222 inhibited proliferation of glioblastoma cells, whereas they had no effect on cell growth when p27 was depleted (le Sage et al. 2007a). These studies established a causal relationship between miR-221/222, p27, and cell proliferation. miR-221/222 also regulate the expression of p27 in prostate carcinoma and melanoma cells, and their over-expression correlates with increased colony-forming potential and proliferation, respectively (Galardi et al. 2007; Felicetti et al. 2008). Collectively, the data suggest that miR-221/222 are oncogenes whose function is to repress the expression of the tumor suppressor p27 (Fig. 4.1).

The translation of p27 is also regulated by other miRNAs in different biological contexts. miR-181a was shown to repress the translation of p27 mRNA in undifferentiated HL60 cells (Cuesta et al. 2009). Repression by miR-181a is relieved during differentiation, allowing the accumulation of p27 necessary to fully block the cell cycle and reach the differentiated state. Intriguingly, one of the two target sites for miR-181a coincides with one of those binding to miR-221/222, suggesting that p27 3′ UTR contains hot-spots for miRNA-mediated regulation.

The modulation of miRNA/target interactions provides an additional level of plasticity to the regulation by miRNAs. Binding of miR-221/222 to their target sites in p27 3′ UTR can be blocked by the RNA-binding protein Dnd1 (Dead end 1), which recognizes U-rich sequences in the vicinity of the miRNA binding sites (Kedde et al. 2007). Dnd1 also counteracts the function of other miRNAs. Thus, cell proliferation and tumor progression should also be influenced by the relative amounts of Dnd1 and miRNAs, at least in primordial germ cells where Dnd1 is expressed.

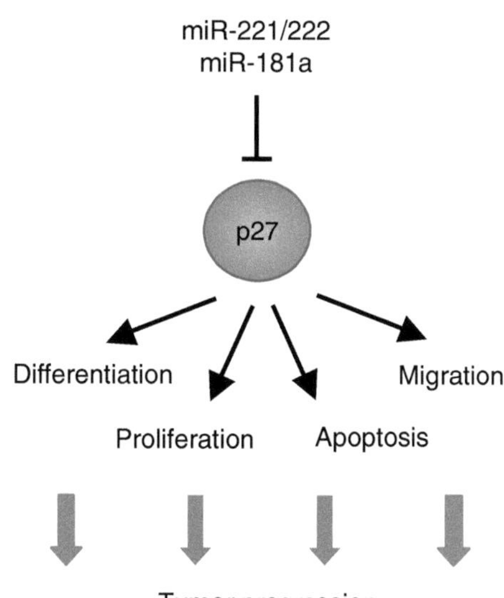

Fig. 4.1 Impact of the translational regulation of p27 mRNA. Translational control of p27 mRNA by miR-221/222 and miR-181a influences cell proliferation, differentiation, and apoptosis. Deregulated expression of miR-221/222 promotes tumor progression (see text for details)

4.5 Conclusions

p27 is a multifunctional protein that performs a dual role: in the nucleus, it acts as a tumor suppressor that inhibits cell proliferation by interfering with the activity of cyclin/CDK complexes; in the cytoplasm, it is an oncogene with prometastatic potential, in part due to its ability to regulate cell migration. Over the years, significant knowledge has accumulated about the mechanisms that regulate p27 protein degradation. Recently, translational regulation of p27 mRNA by miRNAs has emerged as a novel mode of control. Often, miRNAs repress translation only about 2-fold. Since p27 is haploinsufficient for tumor suppression, a reduction of the kind would be enough to promote tumor growth. miR-221 achieves this reduction in a number of tumors and, perhaps for this reason, upregulated miR-221 is part of a miRNA cancer signature. The regulation of p27 mRNA translation and stability are still largely unexplored. Learning about these mechanisms should greatly improve our capacity to develop successful therapies against cancer.

Acknowledgments Work in the author's laboratory was supported by grants 02/028-00 from La Caixa Foundation, grants BMC2003-04108 and BFU2006-01874 from the Spanish Ministry of Education and Science, and grant 2005SGR00669 from the Department of Universities, Information and Sciences of the Generalitat of Catalunya (DURSI).

References

Adams BD, Furneaux H, White BA (2007) The micro-ribonucleic acid (miRNA) miR-206 targets the human estrogen receptor-alpha (ERalpha) and represses ERalpha messenger RNA and protein expression in breast cancer cell lines. Mol Endocrinol 21:1132–47

Aleem E, Kiyokawa H, Kaldis P (2005) Cdc2-cyclin E complexes regulate the G1/S phase transition. Nat Cell Biol 7:831–836

Asangani IA, Rasheed SA, Nikolova DA, Leupold JH, Colburn NH, Post S, Allgayer H (2008) MicroRNA-21 (miR-21) post-transcriptionally downregulates tumor suppressor Pdcd4 and stimulates invasion, intravasation and metastasis in colorectal cancer. Oncogene 27:2128–36

Baldassarre G, Barone MV, Belletti B, Sandomenico C, Bruni P, Spiezia S, Boccia A, Vento MT, Romano A, Pepe S, Fusco A, Viglietto G (1999) Key role of the cyclin-dependent kinase inhibitor p27kip1 for embryonal carcinoma cell survival and differentiation. Oncogene 18:6241–6251

Besson A, Gurian-West M, Schmidt A, Hall A, Roberts JM (2004) p27Kip1 modulates cell migration through the regulation of RhoA activation. Genes Dev 18:862–876

Besson A, Dowdy SF, Roberts JM (2008) CDK inhibitors: cell cycle regulators and beyond. Dev Cell 14:159–169

Blain SW, Scher HI, Cordon-Cardo C, Koff A (2003) p27 as a target for cancer therapeutics. Cancer Cell 3:111–115

Bommer GT, Gerin I, Feng Y, Kaczorowski AJ, Kuick R, Love RE, Zhai Y, Giordano TJ, Clin ZS, Moore BB, MacDongald OA, Cho KR, Fearon ER (2007) P53-Mediated activation of miRNA34 Candidate tumor-suppressor genes. Curr Biol 17:1298–307

Borriello A, Cucciolla V, Oliva A, Zappia V, Della Ragione F (2007) p27Kip1 metabolism: a fascinating labyrinth. Cell Cycle 6:1053–1061

Bryja V, Pacherník J, Soucek K, Horvath V, Dvorák P, Hampl A (2004) Increased apoptosis in differentiating p27-deficient mouse embryonic stem cells. Cell Mol Life Sci 61:1384–1400

Bueno MJ, de Castro IP, Malumbres M (2008) Control of cell proliferation pathways by microR-NAs. Cell Cycle 7:3143–3148

Bushati N, Cohen SM (2007) microRNA functions. Annu Rev Cell Dev Biol 23:175–205

Calin GA, Sevignani C, Dumitru CD, Hyslop T, Noch E, Yendamuri S, Shimizu M, Rattan S, Bullrich F, Negrini M, Croce CM (2004) Human microRNA genes are frequently located at fragile sites and genomic regions involved in cancers. Proc Natl Acad Sci U S A 101:2999–3004

Cho S, Kim JH, Back SH, Jang SK (2005) Polypyrimidine tract-binding protein enhances the internal ribosomal entry site-dependent translation of p27Kip1 mRNA and modulates transition from G1 to S phase. Mol Cell Biol 25:1283–1297

Chu IM, Hengst L, Slingerland JM (2008) The Cdk inhibitor p27 in human cancer: prognostic potential and relevance to anticancer therapy. Nat Rev Cancer 8:253–267

Cimmino A, Calin GA, Fabbri M, Iorio MV, Ferracin M, Shimizu M, Wojcik SE, Aqeilan RI, Zupo S, Dono M, Rassenti L, Alder H, Volinia S, Liu CG, Kipps TJ, Negrini M, Croce CM (2005) miR-15 and miR-16 induce apoptosis by targeting BCL2. Proc Natl Acad Sci U S A 102:13944–13949

Costinean S, Zanesi N, Pekarsky Y, Tili E, Volinia S, Heerema N, Croce CM (2006) Pre-B cell proliferation and lymphoblastic leukemia/high-grade lymphoma in E(mu)-miR155 transgenic mice. Proc Natl Acad Sci U S A 103:7024–7029

Cuesta R, Martínez-Sánchez A, Gebauer F (2009) miR-181a regulates cap-dependent translation of p27[kip1] mRNA in myeloid cells. Mol Cell Biol 29: 2841–2851

Dellas A, Schultheiss E, Leivas MR, Moch H, Torhorst J (1998) Association of p27Kip1, cyclin E and c-myc expression with progression and prognosis in HPV-positive cervical neoplasms. Anticancer Res 18:3991–3998

Denicourt C, Saenz CC, Datnow B, Cui XS, Dowdy SF (2007) Relocalized p27Kip1 tumor suppressor functions as a cytoplasmic metastatic oncogene in melanoma. Cancer Res 67:9238–9243

Eulalio A, Huntzinger E, Izaurralde E (2008) Getting to the root of miRNA-mediated gene silencing. Cell 132:9–14

Felicetti F, Errico MC, Bottero L, Segnalini P, Stoppacciaro A, Biffoni M, Felli N, Mattia G, Petrini M, Colombo MP, Peschle C, Carè A (2008) The promyelocytic leukemia zinc finger-microRNA-221/-222 pathway controls melanoma progression through multiple oncogenic mechanisms. Cancer Res 68:2745–2754

Fero ML, Rivkin M, Tasch M, Porter P, Carow CE, Firpo E, Polyak K, Tsai LH, Broudy V, Perlmutter RM, Kaushansky K, Roberts JM (1996) A syndrome of multiorgan hyperplasia with features of gigantism, tumorigenesis, and female sterility in p27(Kip1)-deficient mice. Cell 85:733–744

Fero ML, Randel E, Gurley KE, Roberts JM, Kemp CJ (1998) The murine gene p27Kip1 is haplo-insufficient for tumour suppression. Nature 396:177–180

Filipowicz W, Bhattacharyya SN, Sonenberg N (2008) Mechanisms of post-transcriptional regulation by microRNAs: are the answers in sight? Nat Rev Genet 9:102–114

Fontana L, Fiori ME, Albini S, Cifaldi L, Giovinazzi S, Forloni M, Boldrini R, Donfrancesco A, Federici V, Giacomini P, Peschle C, Fruci D (2008) Antagomir-17–5p abolishes the growth of therapy-resistant neuroblastoma through p21 and BIM. PLoS ONE 3:e2236

Galardi S, Mercatelli N, Giorda E, Massalini S, Frajese GV, Ciafrè SA, Farace MG (2007) miR-221 and miR-222 expression affects the proliferation potential of human prostate carcinoma cell lines by targeting p27Kip1. J Biol Chem 282:23716–23724

Gillies JK, Lorimer IA (2007) Regulation of p27Kip1 by miRNA 221/222 in glioblastoma. Cell Cycle 6:2005–2009

Gonzalez T, Seoane M, Caamano P, Vinuela J, Dominguez F, Zalvide J (2003) Inhibition of Cdk4 activity enhances translation of p27kip1 in quiescent Rb-negative cells. J Biol Chem 278:12688–12695

Göpfert U, Kullmann M, Hengst L (2003) Cell cycle-dependent translation of p27 involves a responsive element in its 5'-UTR that overlaps with a uORF. Hum Mol Genet 12:1767–1779

Hatfield SD, Shcherbata HR, Fischer KA, Nakahara K, Carthew RW, Ruohola-Baker H (2005) Stem cell division is regulated by the microRNA pathway. Nature 435:974–978

He L, Thomson JM, Hemann MT, Hernando-Monge E, Mu D, Goodson S, Powers S, Cordon-Cardo C, Lowe SW, Hannon GJ, Hammond SM (2005) A microRNA polycistron as a potential human oncogene. Nature 435:828–833

He L, He X, Lim LP, de Stanchina E, Xuan Z, Liang Y, Xue W, Zender L, Magnus J, Ridzon D, Jackson AL, Linsley PS, Chen C, Lowe SW, Cleary MA, Hannon GJ (2007) A microRNA component of the p53 tumour suppressor network. Nature 28:1130–1134

Hengst L, Reed SI (1996) Translational control of p27Kip1 accumulation during the cell cycle. Science 271:1861–1864

Hossain A, Kuo MT, Saunders GF (2006) Mir-17-5p regulates breast cancer cell proliferation by inhibiting translation of AIB1 mRNA. Mol Cell Biol 26:8191–201

Jackson RJ, Standart N (2007) How do microRNAs regulate gene expression? Sci STKE 367:re1

Jiang H, Coleman J, Miskimins R, Srinivasan R, Miskimins WK (2007) Cap-independent translation through the p27 5'-UTR. Nucl Acids Res 35:4767–4778

Karube Y, Tanaka H, Osada H, Tomida S, Tatematsu Y, Yanagisawa K, Yatabe Y, Takamizawa J, Miyoshi S, Mitsudomi T, Takahashi T (2005) Reduced expression of Dicer associated with poor prognosis in lung cancer patients. Cancer Sci 96:111–115

Kedde M, Strasser MJ, Boldajipour B, Oude Vrielink JA, Slanchev K, le Sage C, Nagel R, Voorhoeve PM, van Duijse J, Ørom UA, Lund AH, Perrakis A, Raz E, Agami R (2007) RNA-binding protein Dnd1 inhibits microRNA access to target mRNA. Cell 131:1273–1286

Kiyokawa H, Kineman RD, Manova-Todorova KO, Soares VC, Hoffman ES, Ono M, Khanam D, Hayday AC, Frohman LA, Koff A (1996) Enhanced growth of mice lacking the cyclin-dependent kinase inhibitor function of p27(Kip1). Cell 85:721–732

Koff A (2006) How to decrease p27Kip1 levels during tumor development. Cancer Cell 9:75–76

Kouvaraki M, Gorgoulis VG, Rassidakis GZ, Liodis P, Markopoulos C, Gogas J, Kittas C (2002) High expression levels of p27 correlate with lymph node status in a subset of advanced invasive breast carcinomas: relation to E-cadherin alterations, proliferative activity, and ploidy of the tumors. Cancer 94:2454–2465

Kullmann M, Gopfert U, Siewe B, Hengst L (2002) ELAV/Hu proteins inhibit p27 translation via an IRES element in the p27 5'UTR. Genes Dev 16:3087–3099

Kumar MS, Lu J, Mercer KL, Golub TR, Jacks T (2007) Impaired microRNA processing enhances cellular transformation and tumorigenesis. Nat Genet 39:673–677

Laneve P, Di Marcotullio L, Gioia U, Fiori ME, Ferretti E, Gulino A, Bozzoni I, Caffarelli E (2007) The interplay between microRNAs and the neurotrophin receptor tropomyosin-related kinase C controls proliferation of human neuroblastoma cells. Proc Natl Acad Sci U S A 104:7957–62

le Sage C, Nagel R, Egan DA, Schrier M, Mesman E, Mangiola A, Anile C, Maira G, Mercatelli N, Ciafrè SA, Farace MG, Agami R (2007a) Regulation of the p27(Kip1) tumor suppressor by miR-221 and miR-222 promotes cancer cell proliferation. EMBO J 26:3699–3708

le Sage C, Nagel R, Agami R (2007b) Diverse ways to control p27Kip1 function: miRNAs come into play. Cell Cycle 6:2742–2749

Lee YS, Dutta A (2009) MicroRNAs in cancer Annu Rev Pathol 4:199–277

Lehmann U, Hasemeier B, Christgen M, Müller M, Römermann D, Länger F, Kreipe H (2008) Epigenetic inactivation of microRNA gene hsa-mir-9–1 in human breast cancer. J Pathol 214:17–24

Liu Z, Dong Z, Han B, Yang Y, Liu Y, Zhang JT (2005) Regulation of expression by promoters versus internal ribosome entry site in the 5'-untranslated sequence of the human cyclin-dependent kinase inhibitor p27kip1. Nucl Acids Res 33:3763–3771

Lo AK, To KF, Lo KW, Lung RW, Hui JW, Liao G, Hayward SD (2007) Modulation of LMP1 protein expression by EBV-encoded microRNAs. Proc Natl Acad Sci U S A 104:16164–16169

Lu Z, Liu M, Stribinskis V, Klinge CM, Ramos KS, Colburn NH, Li Y (2008) MicroRNA-21 promotes cell transformation by targeting the programmed cell death 4 gene. Oncogene 27:4373–4379

Lujambio A, Ropero S, Ballestar E, Fraga MF, Cerrato C, Setién F, Casado S, Suarez-Gauthier A, Sanchez-Cespedes M, Git A, Spiteri I, Das PP, Caldas C, Miska E, Esteller M (2007) Genetic unmasking of an epigenetically silenced microRNA in human cancer cells. Cancer Res 67:1424–1429

Ma L, Teruya-Feldstein J, Weinberg RA (2007) Tumour invasion and metastasis initiated by microRNA-10b in breast cancer. Nature 449:682–688

Malumbres M, Barbacid M (2007) Cell cycle kinases in cancer. Curr Opin Genet Dev 17:60–65

Mayr C, Hemann MT, Bartel DP (2007) Disrupting the pairing between let-7 and Hmga2 enhances oncogenic transformation. Science 315:1576–9

Meng F, Henson R, Wehbe-Janek H, Smith H, Ueno Y, Patel T (2007) The MicroRNA let-7a modulates interleukin-6-dependent STAT-3 survival signaling in malignant human cholangiocytes. J Biol Chem 282:8256–8264

Millard SS, Yan JS, Nguyen H, Pagano M, Kiyokawa H, Koff A (1997) Enhanced ribosomal association of p27(Kip1) mRNA is a mechanism contributing to accumulation during growth arrest. J Biol Chem 272:7093–7098

Millard SS, Vidal A, Markus M, Koff A (2000) A U-rich element in the 5' untranslated region is necessary for the translation of p27 mRNA. Mol Cell Biol 20:5947–5959

Miskimins WK, Wang G, Hawkinson M, Miskimins R (2001) Control of cyclin-dependent kinase inhibitor p27 expression by cap-independent translation. Mol Cell Biol 21:4960–4967

Mott JL, Kobayashi S, Bronk SF, Gores GJ (2007) Mir-29 regulates Mcl-1 protein expression and apoptosis. Oncogene 26:6133–6140

Nakayama K, Ishida N, Shirane M, Inomata A, Inoue T, Shishido N, Horii I, Loh DY, Nakayama K (1996) Mice lacking p27(Kip1) display increased body size, multiple organ hyperplasia, retinal dysplasia, and pituitary tumors. Cell 85:707–720

Nguyen L, Besson A, Heng JI, Schuurmans C, Teboul L, Parras C, Philpott A, Roberts JM, Guillemot F (2006) p27kip1 independently promotes neuronal differentiation and migration in the cerebral cortex. Genes Dev 20:1511–1524

Ørom UA, Nielsen FC, Lund AH (2008) MicroRNA-10a binds the 5'UTR of ribosomal protein mRNAs and enhances their translation. Mol Cell 30:460–471

Philipp-Staheli J, Payne SR, Kemp CJ (2001) p27(Kip1): regulation and function of a haploinsufficient tumor suppressor and its misregulation in cancer. Exp Cell Res 264:148–168

Pierson J, Hostager B, Fan R, Vibhakar R (2008) Regulation of cyclin dependent kinase 6 by microRNA 124 in medulloblastoma. J Neurooncol 90:1–7

Richter JD (2008) Think you know how miRNAs work? Think again. Nat Struct Mol Biol 15:334–336

Russo AA, Jeffrey PD, Patten AK, Massagué J, Pavletich NP (1996) Crystal structure of the p27Kip1 cyclin-dependent-kinase inhibitor bound to the cyclin A-Cdk2 complex. Nature 382:325–331

Shell S, Park SM, Radjabi AR, Schickel R, Kistner EO, Jewell DA, Feig C, Lengyel E, Peter ME (2007) Let-7 expression defines two differentiation stages of cancer. Proc Natl Acad Sci U S A 104:11400–11405

Sherr CJ, Roberts JM (2004) Living with or without cyclins and cyclin-dependent kinases. Genes Dev 18:2699–2711

Standart N, Jackson RJ (2007) MicroRNAs repress translation of m7G ppp-capped target mRNAs in vitro by inhibiting initiation and promoting deadenylation. Genes Dev 21:1975–1982

Sylvestre Y, De Guire V, Querido E, Mukhopadhyay UK, Bourdeau V, Major F, Ferbeyre G, Chartrand P (2007) An E2F/miR-20a autoregulatory feedback loop. J Biol Chem 282:2135–2143

Tagawa H, Karube K, Tsuzuki S, Ohshima K, Seto M (2007) Synergistic action of the microRNA-17 polycistron and Myc in aggressive cancer development. Cancer Sci. 98:1482–90

Takamizawa J, Konishi H, Yanagisawa K, Tomida S, Osada H, Endoh H, Harano T, Yatabe Y, Nagino M, Nimura Y, Mitsudomi T, Takahashi T (2004) Reduced expression of the let-7 microRNAs in human lung cancers in association with shortened postoperative survival. Cancer Res 64:3753–3756

Vervoorts J, Lüscher B (2008) Post-translational regulation of the tumor suppressor p27(KIP1). Cell Mol Life Sci 65:3255–3264

Vidal A, Millard SS, Miller JP, Koff A (2002) Rho activity can alter the translation of p27 mRNA and is important for RasV12-induced transformation in a manner dependent on p27 status. J Biol Chem 277:16433–16440

Voorhoeve PM, le Sage C, Schrier M, Gillis AJ, Stoop H, Nagel R, Liu YP, van Duijse J, Drost J, Griekspoor A, Zlotorynski E, Yabuta N, De Vita G, Nojima H, Lee DY, Deng Z, Wang CH, Yang BB (2007) MicroRNA-378 promotes cell survival, tumor growth, and angiogenesis by targeting SuFu and Fus-1 expression. Proc Natl Acad Sci U S A 104:20350–2035

Vrhovac R, Delmer A, Tang R, Marie JP, Zittoun R, Ajchenbaum-Cymbalista F (1998) Prognostic significance of the cell cycle inhibitor p27Kip1 in chronic B-cell lymphocytic leukemia. Blood 91:4694–4700

Watanabe J, Sato H, Kanai T, Kamata Y, Jobo T, Hata H, Fujisawa T, Ohno E, Kameya T, Kuramoto H (2002) Paradoxical expression of cell cycle inhibitor p27 in endometrioid adenocarcinoma of the uterine corpus – correlation with proliferation and clinicopathological parameters. Br J Cancer 87:81–85

Yanaihara N, Caplen N, Bowman E, Seike M, Kumamoto K, Yi M, Stephens RM, Okamoto A, Yokota J, Tanaka T, Calin GA, Liu CG, Croce CM, Harris CC (2006) Unique microRNA molecular profiles in lung cancer diagnosis and prognosis. Cancer Cell 9:189–198

Zhu S, Si ML, Wu H, Mo YY (2007) MicroRNA-21 targets the tumor suppressor gene tropomyosin 1 (TPM1). J Biol Chem 282:14328–14336

Chapter 5
The Inhibitory Effect of Apolipoprotein B mRNA-Editing Enzyme Catalytic Polypeptide-Like 3G (APOBEC3G) and Its Family Members on the Activity of Cellular MicroRNAs

Hui Zhang

Contents

Abstract The apolipoprotein B mRNA-editing enzyme catalytic polypeptide-like 3G (APOBEC3G or APOBEC3G) and its fellow cytidine deaminase family members are potent restrictive factors for human immunodeficiency virus type 1 (HIV-1) and many other retroviruses. However, the cellular function of APOBEC3G remains to be further clarified. It has been reported that APOBEC3s can restrict the mobility of endogenous retroviruses and LTR-retrotransposons, suggesting that they can maintain stability in host genomes. However, APOBEC3G is normally cytoplasmic. Further studies have demonstrated that it is associated with an RNase-sensitive high molecular mass (HMM) and located in processing bodies (P-bodies) of replicating T-cells, indicating that the major cellular function of APOBEC3G seems to be related to P-body-related RNA processing and metabolism. As the function of P-body is closely related to miRNA activity, APOBEC3G could affect the miRNA function. Recent studies have demonstrated that APOBEC3G and its family members counteract miRNA-mediated repression of protein translation. Further, APOBEC3G enhances the association of miRNA-targeted mRNA with polysomes, and facilitates the dissociation of miRNA-targeted mRNA from P-bodies. As such, APOBEC3G regulate the activity of cellular miRNAs. Whether this function is related to its potent antiviral activity remains to be further determined.

H. Zhang
Center for Human Virology, Thomas Jefferson University,
809 Curtis Building, 1015 Walnut Street, Philadelphia, PA 19107, USA
e-mail: hui.zhang@jefferson.edu

R.E. Rhoads (ed.), *miRNA Regulation of the Translational Machinery*,
Progress in Molecular and Subcellular Biology 50,
DOI 10.1007/978-3-642-03103-8_5, © Springer-Verlag Berlin Heidelberg 2010

5.1   The Antiviral Activity of APOBEC3G

The apolipoprotein A3G mRNA-editing enzyme catalytic polypeptide-like 3G (APOBEC3G or A3G) belongs to an APOBEC3 family that has cytidine deaminase (CD) activity (Jarmuz et al. 2002; Sheehy et al. 2002; Zhang et al. 2003). The APOBEC3 genes are arranged in tandem on chromosome 22 (Jarmuz et al. 2002). The proteins of APOBEC3s contain one or two CD domains with the consensus sequence H–X$_N$–C–X$_{2-4}$–C which binds to zinc (Turelli and Trono 2005). Although both CD domains contribute to the CD activity in an in vitro free molecular system (Zhang et al. 2003), the data from an *E. coli* system indicate that the cytidine deaminase activity of intracellular APOBEC3G is mainly performed by the C-terminalmost CD domain (Bishop et al. 2006; Bogerd et al. 2007; Holmes et al. 2007; Newman et al. 2005). APOBEC3 is closely related to APOBEC1, a CD that causes a specific cytosine to uracil change in apolipoprotein B mRNA, and AID enzyme that causes hypermutation of immunoglobulin genes (Navaratnam and Sarwar 2006). APOBEC3 edits the nucleic acids of various retrotransposons and restricts the mobility of endogenous retroviruses and LTR-retrotransposons (Bogerd et al. 2006; Esnault et al. 2005).

APOBEC3G can be found in various cells, such as H9 T-cells, primary CD4 T-cells, macrophages, and many other normal tissues/organs, such as spleen, thymus, testis, ovary, small intestine, and mucosal lining of the colon (Jarmuz et al. 2002; Sheehy et al. 2002). APOBEC3G and many family fellow members potently inhibit the replication of various retroviruses including human immunodeficiency virus (HIV), simian immunodeficiency virus (SIV), type C retroviruses, and hepatitis B virus (Bishop et al. 2004; Doehle et al. 2005; Esnault et al. 2005; Noguchi et al. 2005; Sheehy et al. 2002; Yu et al. 2004a; Zheng et al. 2004). They can either edit the newly synthesized viral DNA or have inhibitory effects at other site(s) of the viral life cycle (Mangeat et al. 2003; Mariani et al. 2003; Turelli et al. 2004; Zhang et al. 2003). APOBEC3G can be efficiently incorporated into HIV-1 particles and causes extensive cytosine to uracil conversion in the viral minus-stranded DNA during reverse transcription (Mangeat et al. 2003; Yu et al. 2004b; Zhang et al. 2003). C–U conversion in minus-stranded DNA could lead to G–A hypermutation in the plus-stranded DNA. The double-stranded DNA harboring G–A hypermutation would encode viral proteins containing aberrant premature stop codons or mutated proteins and would lead to accumulated damage during viral replication (Yu et al. 2004b). Further, APOBEC3G can directly inhibit reverse transcription at multiple steps including primer tRNA annealing, minus and plus strand transfer, primer tRNA processing and removal, and DNA elongation (Bishop et al. 2008; Guo et al. 2006, 2007; Iwatani et al. 2007; Li et al. 2007).

For survival, retroviruses encode various gene products to counteract the inhibition by CDs. In the case of HIV-1 and many other lentiviruses, virion infectivity factor (Vif) is encoded to effectively counter the antiviral effect of APOBEC3G and APOBEC3F by facilitating the degradation of these CDs (Sheehy et al. 2002; Yu et al. 2003; Zheng et al. 2004). Vif can bind to a ubiquitin E3 ligase complex consisting of Cullin 5, elongin B, elongin C, and Rbx1 subunits. The Vif–elongin BC–Cullin 5 complex catalyzes the polyubiquitination of APOBEC3G or APOBEC3F which leads to their degradation in the proteasome (Ehrlich and Yu 2006; Marin et al. 2003; Mehle et al. 2004b; Sheehy et al. 2003; Wichroski et al. 2005; Yu et al. 2003). It has been proposed that the C-terminus of Vif harbors a conserved SLQ sequence and a proline-enriched sequence that are similar to the SOCS box (Kobayashi et al. 2005; Mehle et al. 2004a; Yu et al. 2004c). This SOCS box could serve as an adaptor to mediate the binding between APOBEC3G and the E3 ligase complex. As wild-type HIV-1 harbors the functional *vif*, APOBEC3G/F in the virus-producing cells can be effectively eliminated.

5.2 The Translational Suppression Function of MicroRNA

MicroRNAs (miRNAs) are 20–22-nt-long RNAs that participate in the regulation of various biological processes in numerous eukaryotic lineages, including plants, insects, vertebrate, and mammals (Ambros and Chen 2007; Bartel 2004). The first miRNAs identified in eukaryotes were *lin-4* and *let-7*, discovered by mapping mutant *C. elegans* loci (Lee et al. 1993; Reinhart et al. 2000). Since then, additional miRNAs have been identified by various assays such as genetic screening or computational prediction supported by experimentation (Ambros et al. 2003; Grad et al. 2003). In humans, more than 851 miRNAs have so far been identified, and approximately 30% of genes have been predicted to be subject to miRNA regulation (Lewis et al. 2005). However, a recent assay (rna22) predicted that 92.3% of 3'-untranslated region (3'-UTRs) of human genes harbor at least one target island (each of which corresponds to at least one putative binding site) (Miranda et al. 2006). The expression of many miRNAs is usually tissue or developmental stage-specific, and changes in miRNA expression patterns can be found in the development of many diseases (Ambros and Chen 2007; Calin and Croce 2006; Farh et al. 2005; Lim et al. 2005).

Double-stranded miRNA is generated from primary miRNA by sequential digestion with a Drosha-DGCR8 complex (DiGeorge syndrome critical region gene-8) and Dicer RNase-III endonucleases (Gregory et al. 2004; Grishok et al. 2001; Han et al. 2004; Hutvagner et al. 2001; Lee et al. 2003). Primary miRNA is transcribed from genomic DNA by RNA polymerase II. Drosha-DGCR8 cleaves at sites near the base of the stem in the primary miRNA structure to generate a 60–70-nt fragment comprising the majority of the hairpin (Han et al. 2006; Lee et al. 2003). Dicer subsequently cleaves at sites near the loop to generate the short, double-stranded miRNAs (Bernstein et al. 2001; Han et al. 2004; Tijsterman and Plasterk 2004). These mature miRNAs are then incorporated into the miRNA-induced silencing complex (miRISC), whose composition is similar to that of the RNA-induced silencing complex (RISC) responsible for messenger RNA (mRNA) cleavage, which is guided by small interfering RNAs (siRNAs) (Rana 2007).

The RISC complex contains many proteins. Dicer and Argonaute2 (Ago2) are the essential components. However, several other proteins, such as Tar-binding protein (TRBP), FMRP, and PACT, for example, have been identified in RISC (Caudy et al. 2002; Chendrimada et al. 2005; Gregory et al. 2005; Lee et al. 2006; Meister et al. 2005). The function of these proteins remains to be further clarified. It is notable that MOV10 has been found in a multiprotein complex containing the antiassociation factor eIF6 (Chendrimada et al. 2007). This complex is associated with RISC. Mov10 homologs Armitage in *Drosophila* and SDE3 in *Arabidopsis*, which are putative RNA helicases, play an important role in RISC function (Cook et al. 2004; Lim and Kai 2007; Tomari et al. 2004). Further, Armitage is involved in degradative control of the RISC pathway in neurons, and underlies the pattern of synaptic protein synthesis associated with stable memory (Ashraf et al. 2006). In human cells, siRNA- or miRNA-mediated RNA cleavage requires Mov10, which has also been found in processing bodies (P-bodies) (Gallois-Montbrun et al. 2007). RISC has been shown to load miRNA or siRNA duplexes, not single strands, onto Ago2. The 5' end of the duplex is lodged in the phosphate-binding pocket of the Ago2 Piwi domain (Ma et al. 2005; Parker et al. 2005). Once bound to the miRNA or siRNA duplex, Ago2 cleaves the passenger strand, triggering its dissociation from the complex and the concomitant maturation of the active RISC (Matranga et al. 2005). Passenger-strand cleavage is not obligatory but is the normal mechanism for RISC activation. In the mature RISC, the guiding-strand of siRNA is associated with Ago2. In humans, Ago2 is the sole siRNA-guided Argonaute protein able to act as an RNA-guided, Mg^{2+}-dependent RNA endonuclease that cleaves a single phosphodiester bond in the target mRNA, triggering its destruction (Liu et al. 2004; Meister et al. 2004).

However, the component of miRISC remains to be further clarified. All four Ago proteins, Ago1–4, can be found in miRISC (Meister et al. 2005). Although the miRNA bound to Ago2 can serve as the guiding RNA for cleavage of the target mRNA (Meister et al. 2004), it mainly serves as a guide to direct the repression of protein synthesis. Ago1, but not Ago2, mediates this process in *Drosophila* (Behm-Ansmant et al. 2006; Forstemann et al. 2007; Okamura et al. 2004; Tomari et al. 2007). Whether mammalian cells also use Ago1 to mediate this process remains to be clarified. The mature miRNAs typically interact with target mRNAs by partial sequence matching. A 7-nt "seed" sequence in the miRNA at position 2–8 from the 5' end seems to be essential for miRNA action in human cells, and the function of the remaining nucleotides appears less critical (Lewis et al. 2005). The miRNA binds to a complementary sequence in the 3'-UTR of its target mRNA, resulting in degradation of the mRNA transcript and/or translational inhibition (Bartel 2004).

The translation inhibition mediated by miRNA is not well understood (Pillai et al. 2007); however, in most cases, it is not the result of miRNA-mediated mRNA degradation (Bartel 2004; Pillai et al. 2007). Some reports indicate that miRNAs interfere with the initiation of protein synthesis (Chendrimada et al. 2007; Eulalio et al. 2007a; Humphreys et al. 2005; Pillai et al. 2005, 2007). Because miRNAs only inhibit cap-dependent translation and not translation resulting from an internal ribosome entry site (IRES), it is possible that miRNAs interfere with $m^{7}G$ cap recognition (Pillai et al. 2005). As the sequences of central domain of Ago protein is similar to eIF4e, an essential factor for cap-dependent translation, Ago proteins could competitively bind to cap structure, thereby inhibiting the initiation of translation (Kiriakidou et al. 2007). Conversely, because miRNAs are associated with the translating mRNA and prevent the generation of the nascent polypeptide in the polysome, it seems that they also directly inhibit protein synthesis at the polysome at the post-initiation stage (Maroney et al. 2006; Nottrott et al. 2006; Olsen and Ambros 1999). It is notable that poly-A binding protein 1 (PABP1), which binds to the poly-A tail at the 3′-end of mRNAs, interacts with the elongation initiation factor 4G (eIF4G), which is a part of the initiation complex eIF4F, to start the assembly of the ribosome at the 5′-end of the mRNA. As a result, the two ends of the mRNA are bound together. The region of the 3′-UTR to which the miRNA binds is quite close to the initiation site of translation (Gingras et al. 1999). miRNA-bound Ago/GW182 could prevent this circularization.

5.3 The Role of P-bodies in Regulating the Function of MicroRNAs

Recent developments have indicated the involvement of P-bodies in miRNA-mediated translation inhibition (Bruno and Wilkinson 2006; Chu and Rana 2006; Eulalio et al. 2007a; Pillai et al. 2007). P-bodies are also named GW or Dcp bodies and are localized to the cytoplasm. P-bodies harbor many proteins involved in RNA processing and degradation, such as GW182, Dcp1/2, Xrn1, Lsm1, RCK/p54, and eIF4E (Eulalio et al. 2007a). For instance, Dcp1 and Dcp2 catalyze the decapping of mRNA, which is followed by the 5′→3′ exonucleolytic degradation of mRNA via Xrn1 (Eulalio et al. 2007a; Pillai et al. 2007). The formation of a P-body is a dynamic process and requires a consistent accumulation of miRNA-repressed mRNAs (Eulalio et al. 2007a). Several lines of evidence have indicated that P-bodies are actively involved in miRNA-mediated mRNA repression (Eulalio et al. 2007a). The P-body-associated protein GW182 associates directly with Ago-1, even before the formation of a P-body (Behm-Ansmant et al. 2006; Liu et al. 2005). Depletion of P-body components such as GW182 and Rck/p54 prevents translational repression of target mRNAs (Behm-Ansmant et al. 2006; Bruno and Wilkinson 2006; Chu and Rana 2006; Eulalio et al. 2007a; Jakymiw et al. 2005; Liu et al. 2005; Pillai et al. 2007). Furthermore, several miRISC-related components, such as miRNAs, mRNAs repressed by miRNAs, Ago-1, Ago-2, and Mov10, are found in P-bodies (Bruno and Wilkinson 2006; Eulalio et al. 2007a; Pillai et al. 2007). The formation of P-body is a dynamic process that requires continuous accumulation of repressed mRNAs (Eulalio et al. 2007b).

However, P-bodies also function as storage sites for excess mRNAs (Bruno and Wilkinson 2006; Eulalio et al. 2007a). Stored mRNAs can be sent back to polysomes when further protein synthesis is required. In fact, some cellular proteins can facilitate the exit of miRNA-bound mRNAs from P-bodies. Stress may induce the relocation of HuR, an AU-enriched-element binding protein, from the nucleus to the cytoplasm, where it can bind to the 3′-UTR of target mRNAs (e.g., CAT-1) stored in P-bodies. This binding increases the stability of the miR-122-bound mRNA by assisting it to egress from the P-body and return to polysomes (Bhattacharyya et al. 2006).

5.4 APOBEC3G Counteracts miRNA-Mediated Repression of Protein Translation

Interestingly, APOBEC3G is found in P-bodies and stress granules (Gallois-Montbrun et al. 2007; Wichroski et al. 2006). It is associated with a high molecular mass (HMM) structure (>700 kDa) in replicating cells, and this interaction is RNase-sensitive (Chiu et al. 2005, 2006). We have also found that APOBEC3G can bind with some of these RNA-binding proteins by a tandem affinity purification (TAP) assay (Fig. 5.1) developed by Dr. Bertrand Séraphin's group (Puig et al. 2001; Rigaut et al. 1999). This result clearly indicated that Mov 10 is an APOBEC3G associated protein. Two other studies indicate that APOBEC3G interacts with many RNA-binding proteins, among which are several miRNA-related proteins, such as Ago1, Ago2, Mov10, GW182, and PABP1. These interactions are either partially

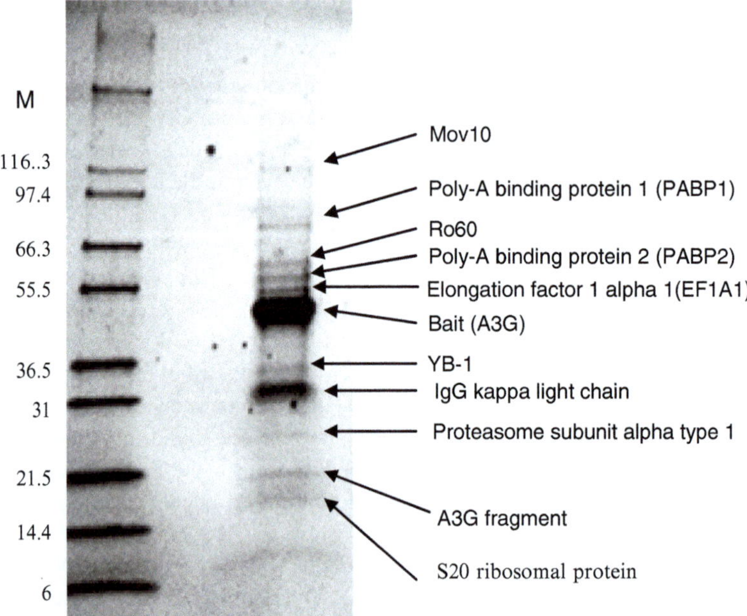

Fig. 5.1 APOBEC3G associated proteins identified with tandem affinity purification (TAP) assay

or completely resistant to RNase A digestion (Chiu et al. 2006; Gallois-Montbrun et al. 2007; Kozak et al. 2006). Therefore, aside from its inhibitory function in relation to endogenous retroviruses and other retrotransposons (Bogerd et al. 2006; Dutko et al. 2005; Esnault et al. 2005, 2006; Hulme et al. 2007), the major cellular function of APOBEC3G seems to be related to miRNA activity and P-body related RNA processing and metabolism.

To study whether APOBEC3G affects the efficiency of miRNA-mediated translational repression, various 293T-cell-enriched miRNA-binding sites with perfect or partial complementarity to their corresponding miRNAs were inserted into the 3′-UTR of luciferase (luc) or *gfp* (Huang et al. 2007). These plasmids were transfected into 293T-cells, which naturally do not express APOBEC3G (Sheehy et al. 2002; Yang et al. 2007), with or without an APOBEC3G-HA-expressing plasmid. Our studies have indicated that APOBEC3G significantly counteracted the inhibitory effect of 3′-UTR miRNA binding site on the expression of reporter gene. Similar phenomenon can be observed in HeLa cells. Conversely, inhibition of APOBEC3G expression in some APOBEC3G- and APOBEC3F-enriched cells, such as H9 T-cells, PHA-activated primary CD4+ T-lymphocytes, and macrophages, with APOBEC3G- and APOBEC3F-specific siRNAs will enhance the expression of reporter genes which harbor 3′-UTR miRNA binding sites. In addition, many other APOBEC3 family members such as A3B, A3C, and A3F are able to inhibit the miRNA-mediated translational repression. Moreover, the mutations that inactivate the N-terminal domain, C97A and C100A, had a modest effect on miRNA-mediated translational repression, whereas the C-terminal domain C288A and C291A mutations had no significant influence on the inhibitory effect of APOBEC3G, suggesting that the CD activity is unlikely involved in this inhibitory effect (Huang et al. 2007).

Further studies on mechanism have demonstrated that APOBEC3G significantly enhance the association of the target mRNA with polysomes. Conversely, APOBEC3G decreases the association of the target mRNA with P-bodies. Immunoprecipitation and confocal studies further indicate that APOBEC3G interacts with GW182 (Huang et al. 2007). Moreover, the depletion of GW182 with GW182-specific siRNA had a synergistic effect with APOBEC3G in counteracting miRNA-mediated translational repression, which is consistent with previous reports regarding the role of GW182 in miRNA function (Behm-Ansmant et al. 2006; Liu et al. 2005).

Given that APOBEC3G is associated with mRNA, localizes to P-bodies and stress granules (Gallois-Montbrun et al. 2007; Kozak et al. 2006; Wichroski et al. 2006), and can substantially enhance the expression of miRNA-targeted mRNA, it is unlikely that APOBEC3G directly improves the interaction between mRNA and polysomes or inhibits the interaction between miRNA and its target mRNA in miRISC. Instead, APOBEC3G may directly counteracts translational repression mediated by Ago and GW182, may block miRNA-targeted mRNA from entering P-bodies or stress granules, may prevent the miRNA-targeted mRNA from engaging the RNA degradation machinery in P-bodies, or may directly facilitate the egress of miRNA-targeted mRNA from P-bodies and stress granules (Fig. 5.2). By one or more of these approaches, APOBEC3G may inhibit the

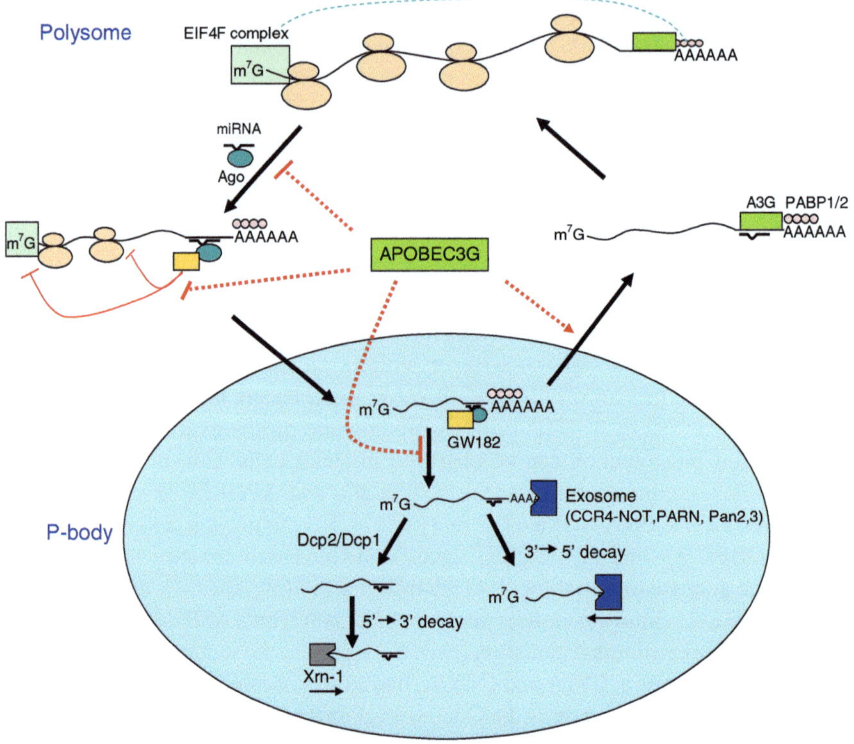

Fig. 5.2 The possible target sites for APOBEC3G to inhibit the activity of miRNAs

degradation or storage of miRNA-targeted miRNA in P-bodies and stress granules. Subsequently, more mRNA could associate with polysomes, and the translation efficiency would therefore be enhanced. However, as the mechanism of the regulation of mRNA degradation and storage in P-bodies or stress granules remains to be clarified and the relationship between miRNA-mediated translational repression and P-bodies is still under intensive investigation, further experiments are required to demonstrate the exact mechanism underlying this cellular function of APOBEC3G.

5.5   Conclusion Remarks

Recently, we and others have found that interferon (IFN)-α/β can significantly enhance the expression of APOBEC3G/F in various primary cells, such as resting CD4 T-lymphocytes, macrophages, endothelial cells, hepatocytes, myeloid dendritic cells, and plasmacytoid dendritic cells (our unpublished data) (Argyris et al. 2007;

Bonvin et al. 2006; Chen et al. 2006; Peng et al. 2006; Sarkis et al. 2006; Tanaka et al. 2006). Therefore, the correlation of IFN regulatory system and the miRNA activity in these primary cells merits to being further investigated. Interestingly, the mutations C228A and C291A inactivated the CD activity of APOBEC3G, but APOBEC3G was still able to enhance the expression of luciferase when *luc* was controlled by miRNA (Fig. 3b). Therefore, the derepression of miRNA-mediated inhibition of protein translation by APOBEC3G is separable from its CD activity. As described in many reports, the CD activity of APOBEC3G is only partially responsible for viral infectivity (Bishop et al. 2006; Bogerd et al. 2007; Holmes et al. 2007; Newman et al. 2005; Zhang et al. 2003). It remains to be determined whether this cellular function of APOBEC3G in protein translation regulation is related to its CD-independent antiviral activity.

Acknowledgments   This work was supported in part by NIH grants AI058798 and AI052732 to H.Z.

References

Ambros V, Chen X (2007) The regulation of genes and genomes by small RNAs. Development 134:1635–1641

Ambros V, Lee RC, Lavanway A, Williams PT, Jewell D (2003) MicroRNAs and other tiny endogenous RNAs in C. elegans. Curr Biol 13:807–818

Argyris EG, Acheampong E, Wang F, Huang J, Chen K, et al (2007) The interferon-induced expression of APOBEC3G in human blood-brain barrier exerts a potent intrinsic immunity to block HIV-1 entry to central nervous system. Virology 367:440–451

Ashraf SI, McLoon AL, Sclarsic SM, Kunes S (2006) Synaptic protein synthesis associated with memory is regulated by the RISC pathway in Drosophila. Cell 124:191–205

Bartel DP (2004) MicroRNAs: genomics, biogenesis, mechanism, and function. Cell 116:281–297

Behm-Ansmant I, Rehwinkel J, Doerks T, Stark A, Bork P, Izaurralde E (2006) mRNA degradation by miRNAs and GW182 requires both CCR4:NOT deadenylase and DCP1:DCP2 decapping complexes. Genes Dev 20:1885–1898

Bernstein E, Caudy AA, Hammond SM, Hannon GJ (2001) Role for a bidentate ribonuclease in the initiation step of RNA interference. Nature 409:363–366

Bhattacharyya SN, Habermacher R, Martine U, Closs EI, Filipowicz W (2006) Relief of microRNA-mediated translational repression in human cells subjected to stress. Cell 125:1111–1124

Bishop KN, Holmes RK, Malim MH (2006) Antiviral potency of APOBEC proteins does not correlate with cytidine deamination. J Virol 80:8450–8458

Bishop KN, Holmes RK, Sheehy AM, Davidson NO, Cho SJ, Malim MH (2004) Cytidine deamination of retroviral DNA by diverse APOBEC proteins. Curr Biol 14:1392–1396

Bishop KN, Verma M, Kim EY, Wolinsky SM, Malim MH (2008) APOBEC3G inhibits elongation of HIV-1 reverse transcripts. PLoS Pathog 4:e1000231

Bogerd HP, Wiegand HL, Doehle BP, Cullen BR. 2007. The intrinsic antiretroviral factor APOBEC3B contains two enzymatically active cytidine deaminase domains. Virology 364:486–493

Bogerd HP, Wiegand HL, Doehle BP, Lueders KK, Cullen BR (2006) APOBEC3A and APOBEC3B are potent inhibitors of LTR-retrotransposon function in human cells. Nucl Acids Res 34:89–95

Bonvin M, Achermann F, Greeve I, Stroka D, Keogh A et al (2006) Interferon-inducible expression of APOBEC3 editing enzymes in human hepatocytes and inhibition of hepatitis B virus replication. Hepatology 43:1364–1374

Bruno I, Wilkinson MF (2006) P-bodies react to stress and nonsense. Cell 125:1036–1038

Calin GA, Croce CM (2006) MicroRNA signatures in human cancers. Nat Rev Cancer 6:857–866

Caudy AA, Myers M, Hannon GJ, Hammond SM (2002) Fragile X-related protein and VIG associate with the RNA interference machinery. Genes Dev 16:2491–2496

Chen K, Huang J, Zhang C, Huang S, Nunnari G et al (2006) Alpha interferon potently enhances the anti-human immunodeficiency virus type 1 activity of APOBEC3G in resting primary CD4 T cells. J Virol 80:7645–7657

Chendrimada TP, Finn KJ, Ji X, Baillat D, Gregory RI et al (2007) MicroRNA silencing through RISC recruitment of eIF6. Nature 447:823–828

Chendrimada TP, Gregory RI, Kumaraswamy E, Norman J, Cooch N et al (2005) TRBP recruits the Dicer complex to Ago2 for microRNA processing and gene silencing. Nature 436:740–744

Chiu YL, Soros VB, Kreisberg JF, Stopak K, Yonemoto W, Greene WC (2005) Cellular APOBEC3G restricts HIV-1 infection in resting CD4+ T cells. Nature 435:108–114

Chiu YL, Witkowska HE, Hall SC, Santiago M, Soros VB et al (2006) High-molecular-mass APOBEC3G complexes restrict Alu retrotransposition. Proc Natl Acad Sci U S A 103:15588–15593

Chu CY, Rana TM (2006) Translation repression in human cells by microRNA-induced gene silencing requires RCK/p54. PLoS Biol 4:e210

Cook HA, Koppetsch BS, Wu J, Theurkauf WE (2004) The Drosophila SDE3 homolog armitage is required for oskar mRNA silencing and embryonic axis specification. Cell 116:817–829

Doehle BP, Schafer A, Wiegand HL, Bogerd HP, Cullen BR (2005) Differential sensitivity of murine leukemia virus to APOBEC3-mediated inhibition is governed by virion exclusion. J Virol 79:8201–8207

Dutko JA, Schafer A, Kenny AE, Cullen BR, Curcio MJ (2005) Inhibition of a yeast LTR retrotransposon by human APOBEC3 cytidine deaminases. Curr Biol 15:661–666

Ehrlich ES, Yu XF (2006) Lentiviral Vif: viral hijacker of the ubiquitin-proteasome system. Int J Hematol 83:208–212

Esnault C, Heidmann O, Delebecque F, Dewannieux M, Ribet D et al (2005) APOBEC3G cytidine deaminase inhibits retrotransposition of endogenous retroviruses. Nature 433:430–433

Esnault C, Millet J, Schwartz O, Heidmann T (2006) Dual inhibitory effects of APOBEC family proteins on retrotransposition of mammalian endogenous retroviruses. Nucl Acids Res 34:1522–1531

Eulalio A, Behm-Ansmant I, Izaurralde E (2007a) P bodies: at the crossroads of post-transcriptional pathways. Nat Rev Mol Cell Biol 8:9–22

Eulalio A, Behm-Ansmant I, Schweizer D, Izaurralde E (2007b) P-body formation is a consequence, not the cause, of RNA-mediated gene silencing. Mol Cell Biol 27:3970–3981

Farh KK, Grimson A, Jan C, Lewis BP, Johnston WK et al (2005) The widespread impact of mammalian MicroRNAs on mRNA repression and evolution. Science 310:1817–1821

Forstemann K, Horwich MD, Wee L, Tomari Y, Zamore PD (2007) Drosophila microRNAs are sorted into functionally distinct argonaute complexes after production by dicer-1. Cell 130:287–297

Gallois-Montbrun S, Kramer B, Swanson CM, Byers H, Lynham S et al (2007) Antiviral protein APOBEC3G localizes to ribonucleoprotein complexes found in P bodies and stress granules. J Virol 81:2165–2178

Gingras AC, Raught B, Sonenberg N (1999) eIF4 initiation factors: effectors of mRNA recruitment to ribosomes and regulators of translation. Annu Rev Biochem 68:913–963

Grad Y, Aach J, Hayes GD, Reinhart BJ, Church GM et al (2003) Computational and experimental identification of C. elegans microRNAs. Mol Cell 11:1253–1263

Gregory RI, Chendrimada TP, Cooch N, Shiekhattar R (2005) Human RISC couples microRNA biogenesis and posttranscriptional gene silencing. Cell 123:631–640

Gregory RI, Yan KP, Amuthan G, Chendrimada T, Doratotaj B et al (2004) The Microprocessor complex mediates the genesis of microRNAs. Nature 432:235–240

Grishok A, Pasquinelli AE, Conte D, Li N, Parrish S et al (2001) Genes and mechanisms related to RNA interference regulate expression of the small temporal RNAs that control C. elegans developmental timing. Cell 106:23–34

Guo F, Cen S, Niu M, Saadatmand J, Kleiman L (2006) Inhibition of formula-primed reverse transcription by human APOBEC3G during human immunodeficiency virus type 1 replication. J Virol 80:11710–11722

Guo F, Cen S, Niu M, Yang Y, Gorelick RJ, Kleiman L (2007) The interaction of APOBEC3G with human immunodeficiency virus type 1 nucleocapsid inhibits tRNA3Lys annealing to viral RNA. J Virol 81:11322–11331

Han J, Lee Y, Yeom KH, Kim YK, Jin H, Kim VN (2004) The Drosha-DGCR8 complex in primary microRNA processing. Genes Dev 18:3016–3027

Han J, Lee Y, Yeom KH, Nam JW, Heo I et al (2006) Molecular basis for the recognition of primary microRNAs by the Drosha-DGCR8 complex. Cell 125:887–901

Holmes RK, Koning FA, Bishop KN, Malim MH (2007) APOBEC3F can inhibit the accumulation of HIV-1 reverse transcription products in the absence of hypermutation. Comparisons with APOBEC3G. J Biol Chem 282:2587–2595

Huang J, Liang Z, Yang B, Tian H, Ma J, Zhang H (2007) Derepression of microRNA-mediated protein translation inhibition by apolipoprotein B mRNA-editing enzyme catalytic polypeptide-like 3G (APOBEC3G) and its family members. J Biol Chem 282:33632–33640

Hulme AE, Bogerd HP, Cullen BR, Moran JV (2007) Selective inhibition of Alu retrotransposition by APOBEC3G. Gene 390:199–205

Humphreys DT, Westman BJ, Martin DI, Preiss T (2005) MicroRNAs control translation initiation by inhibiting eukaryotic initiation factor 4E/cap and poly(A) tail function. Proc Natl Acad Sci U S A 102:16961–16966

Hutvagner G, McLachlan J, Pasquinelli AE, Balint E, Tuschl T, Zamore PD (2001) A cellular function for the RNA-interference enzyme Dicer in the maturation of the let-7 small temporal RNA. Science 293:834–838

Iwatani Y, Chan DS, Wang F, Maynard KS, Sugiura W et al (2007) Deaminase-independent inhibition of HIV-1 reverse transcription by APOBEC3G. Nucl Acids Res 35:7096–7108

Jakymiw A, Lian S, Eystathioy T, Li S, Satoh M et al (2005) Disruption of GW bodies impairs mammalian RNA interference. Nat Cell Biol 7:1267–1274

Jarmuz A, Chester A, Bayliss J, Gisbourne J, Dunham I et al (2002) An anthropoid-specific locus of orphan C to U RNA-editing enzymes on chromosome 22. Genomics 79:285–296

Kiriakidou M, Tan GS, Lamprinaki S, De Planell-Saguer M, Nelson PT, Mourelatos Z (2007) An mRNA m7G cap binding-like motif within human Ago2 represses translation. Cell 129:1141–1151

Kobayashi M, Takaori-Kondo A, Miyauchi Y, Iwai K, Uchiyama T (2005) Ubiquitination of APOBEC3G by an HIV-1 Vif-Cullin5-Elongin B-Elongin C complex is essential for Vif function. J Biol Chem 280:18573–18578

Kozak SL, Marin M, Rose KM, Bystrom C, Kabat D (2006) The anti-HIV-1 editing enzyme APOBEC3G binds HIV-1 RNA and messenger RNAs that shuttle between polysomes and stress granules. J Biol Chem 281:29105–29119

Lee RC, Feinbaum RL, Ambros V (1993) The C. elegans heterochronic gene lin-4 encodes small RNAs with antisense complementarity to lin-14. Cell 75:843–854

Lee Y, Ahn C, Han J, Choi H, Kim J et al (2003) The nuclear RNase III Drosha initiates microRNA processing. Nature 425:415–419

Lee Y, Hur I, Park SY, Kim YK, Suh MR, Kim VN (2006) The role of PACT in the RNA silencing pathway. EMBO J 25:522–532

Lewis BP, Burge CB, Bartel DP (2005) Conserved seed pairing, often flanked by adenosines, indicates that thousands of human genes are microRNA targets. Cell 120:15–20

Li XY, Guo F, Zhang L, Kleiman L, Cen S (2007) APOBEC3G inhibits DNA strand transfer during HIV-1 reverse transcription. J Biol Chem 282:32065–32074

Lim AK, Kai T (2007) Unique germ-line organelle, nuage, functions to repress selfish genetic elements in Drosophila melanogaster. Proc Natl Acad Sci U S A 104:6714–6719

Lim LP, Lau NC, Garrett-Engele P, Grimson A, Schelter JM et al (2005) Microarray analysis shows that some microRNAs downregulate large numbers of target mRNAs. Nature 433:769–773

Liu J, Carmell MA, Rivas FV, Marsden CG, Thomson JM et al (2004) Argonaute2 is the catalytic engine of mammalian RNAi. Science 305:1437–1441

Liu J, Rivas FV, Wohlschlegel J, Yates JR 3rd, Parker R, Hannon GJ (2005) A role for the P-body component GW182 in microRNA function. Nat Cell Biol 7:1261–1266

Ma JB, Yuan YR, Meister G, Pei Y, Tuschl T, Patel DJ (2005) Structural basis for 5'-end-specific recognition of guide RNA by the A. fulgidus Piwi protein. Nature 434:666–670

Mangeat B, Turelli P, Caron G, Friedli M, Perrin L, Trono D (2003) Broad antiretroviral defence by human APOBEC3G through lethal editing of nascent reverse transcripts. Nature 424:99–103

Mariani R, Chen D, Schrofelbauer B, Navarro F, Konig R et al (2003) Species-specific exclusion of APOBEC3G from HIV-1 virions by Vif. Cell 114:21–31

Marin M, Rose KM, Kozak SL, Kabat D (2003) HIV-1 Vif protein binds the editing enzyme APOBEC3G and induces its degradation. Nat Med 9:1398–1403

Maroney PA, Yu Y, Fisher J, Nilsen TW (2006) Evidence that microRNAs are associated with translating messenger RNAs in human cells. Nat Struct Mol Biol 13:1102–1107

Matranga C, Tomari Y, Shin C, Bartel DP, Zamore PD (2005) Passenger-strand cleavage facilitates assembly of siRNA into Ago2-containing RNAi enzyme complexes. Cell 123:607–620

Mehle A, Goncalves J, Santa-Marta M, McPike M, Gabuzda D (2004a) Phosphorylation of a novel SOCS-box regulates assembly of the HIV-1 Vif-Cul5 complex that promotes APOBEC3G degradation. Genes Dev 18:2861–2866

Mehle A, Strack B, Ancuta P, Zhang C, McPike M, Gabuzda D (2004b) Vif overcomes the innate antiviral activity of APOBEC3G by promoting its degradation in the ubiquitin-proteasome pathway. J Biol Chem 279:7792–7798

Meister G, Landthaler M, Patkaniowska A, Dorsett Y, Teng G, Tuschl T (2004) Human Argonaute2 mediates RNA cleavage targeted by miRNAs and siRNAs. Mol Cell 15:185–197

Meister G, Landthaler M, Peters L, Chen PY, Urlaub H et al (2005) Identification of novel argonaute-associated proteins. Curr Biol 15:2149–2155

Miranda KC, Huynh T, Tay Y, Ang YS, Tam WL et al (2006) A pattern-based method for the identification of MicroRNA binding sites and their corresponding heteroduplexes. Cell 126:1203–1217

Navaratnam N, Sarwar R (2006) An overview of cytidine deaminases. Int J Hematol 83:195–200

Newman EN, Holmes RK, Craig HM, Klein KC, Lingappa JR et al (2005) Antiviral function of APOBEC3G can be dissociated from cytidine deaminase activity. Curr Biol 15:166–170

Noguchi C, Ishino H, Tsuge M, Fujimoto Y, Imamura M et al (2005) G to A hypermutation of hepatitis B virus. Hepatology 41:626–633

Nottrott S, Simard MJ, Richter JD (2006) Human let-7a miRNA blocks protein production on actively translating polyribosomes. Nat Struct Mol Biol 13:1108–1114

Okamura K, Ishizuka A, Siomi H, Siomi MC (2004) Distinct roles for Argonaute proteins in small RNA-directed RNA cleavage pathways. Genes Dev 18:1655–1666

Olsen PH, Ambros V (1999) The lin-4 regulatory RNA controls developmental timing in Caenorhabditis elegans by blocking LIN-14 protein synthesis after the initiation of translation. Dev Biol 216:671–680

Parker JS, Roe SM, Barford D (2005) Structural insights into mRNA recognition from a PIWI domain-siRNA guide complex. Nature 434:663–666

Peng G, Lei KJ, Jin W, Greenwell-Wild T, Wahl SM (2006) Induction of APOBEC3 family proteins, a defensive maneuver underlying interferon-induced anti-HIV-1 activity. J Exp Med 203:41–46

Pillai RS, Bhattacharyya SN, Artus CG, Zoller T, Cougot N et al (2005) Inhibition of translational initiation by Let-7 MicroRNA in human cells. Science 309:1573–1576

Pillai RS, Bhattacharyya SN, Filipowicz W (2007) Repression of protein synthesis by miRNAs: how many mechanisms? Trends Cell Biol 17:118–126

Puig O, Caspary F, Rigaut G, Rutz B, Bouveret E et al (2001) The tandem affinity purification (TAP) method: a general procedure of protein complex purification. Methods 24:218–229

Rana TM (2007) Illuminating the silence: understanding the structure and function of small RNAs. Nat Rev Mol Cell Biol 8:23–36

Reinhart BJ, Slack FJ, Basson M, Pasquinelli AE, Bettinger JC et al (2000) The 21-nucleotide let-7 RNA regulates developmental timing in Caenorhabditis elegans. Nature 403:901–906

Rigaut G, Shevchenko A, Rutz B, Wilm M, Mann M, Seraphin B (1999) A generic protein purification method for protein complex characterization and proteome exploration. Nat Biotechnol 17:1030–1032

Sarkis PT, Ying S, Xu R, Yu XF (2006) STAT1-independent cell type-specific regulation of antiviral APOBEC3G by IFN-alpha. J Immunol 177:4530–4540

Sheehy AM, Gaddis NC, Choi JD, Malim MH (2002) Isolation of a human gene that inhibits HIV-1 infection and is suppressed by the viral Vif protein. Nature 418:646–650

Sheehy AM, Gaddis NC, Malim MH (2003) The antiretroviral enzyme APOBEC3G is degraded by the proteasome in response to HIV-1 Vif. Nat Med 9:1404–1407

Tanaka Y, Marusawa H, Seno H, Matsumoto Y, Ueda Y et al (2006) Anti-viral protein APOBEC3G is induced by interferon-alpha stimulation in human hepatocytes. Biochem Biophys Res Commun 341:314–319

Tijsterman M, Plasterk RH (2004) Dicers at RISC; the mechanism of RNAi. Cell 117:1–3

Tomari Y, Du T, Haley B, Schwarz DS, Bennett R et al (2004) RISC assembly defects in the Drosophila RNAi mutant armitage. Cell 116:831–841

Tomari Y, Du T, Zamore PD (2007) Sorting of Drosophila small silencing RNAs. Cell 130:299–308

Turelli P, Mangeat B, Jost S, Vianin S, Trono D (2004) Inhibition of hepatitis B virus replication by APOBEC3G. Science 303:1829

Turelli P, Trono D (2005) Editing at the crossroad of innate and adaptive immunity. Science 307:1061–1065

Wichroski MJ, Ichiyama K, Rana TM (2005) Analysis of HIV-1 viral infectivity factor-mediated proteasome-dependent depletion of APOBEC3G: correlating function and subcellular localization. J Biol Chem 280:8387–8396

Wichroski MJ, Robb GB, Rana TM (2006) Human retroviral host restriction factors APOBEC3G and APOBEC3F localize to mRNA processing bodies. PLoS Pathog 2:e41

Yang B, Chen K, Zhang C, Huang S, Zhang H (2007) Virion-associated Uracil DNA Glycosylase-2 and Apurinic/Apyrimidinic Endonuclease Are Involved in the Degradation of APOBEC3G-edited Nascent HIV-1 DNA. J Biol Chem 282:11667–11675

Yu Q, Chen D, Konig R, Mariani R, Unutmaz D, Landau NR (2004a) APOBEC3B and APOBEC3C are potent inhibitors of simian immunodeficiency virus replication. J Biol Chem 279:53379–53386

Yu Q, Konig R, Pillai S, Chiles K, Kearney M et al (2004b) Single-strand specificity of APOBEC3G accounts for minus-strand deamination of the HIV genome. Nat Struct Mol Biol 11:435–442

Yu X, Yu Y, Liu B, Luo K, Kong W et al (2003) Induction of APOBEC3G ubiquitination and degradation by an HIV-1 Vif-Cul5-SCF complex. Science 302:1056–1060

Yu Y, Xiao Z, Ehrlich ES, Yu X, Yu XF (2004c) Selective assembly of HIV-1 Vif-Cul5-ElonginB-ElonginC E3 ubiquitin ligase complex through a novel SOCS box and upstream cysteines. Genes Dev 18:2867–2872

Zhang H, Yang B, Pomerantz RJ, Zhang C, Arunachalam SC, Gao L (2003) The cytidine deaminase CEM15 induces hypermutation in newly synthesized HIV-1 DNA. Nature 424:94–98

Zheng YH, Irwin D, Kurosu T, Tokunaga K, Sata T, Peterlin BM (2004) Human APOBEC3F is another host factor that blocks human immunodeficiency virus type 1 replication. J Virol 78:6073–6076

Chapter 6
MicroRNA-Mediated mRNA Deadenylation and Repression of Protein Synthesis in a Mammalian Cell-Free System

Motoaki Wakiyama and Shigeyuki Yokoyama

Contents

Abstract Cell-free systems are valuable tools for analyses of a post-transcriptional gene expression. The biochemical aspects of RNA interference have been extensively studied by using extracts prepared from *Drosophila* embryos. However, the mechanism by which microRNAs regulate protein synthesis is still elusive. We established a mammalian cell-free system that recapitulates *let-7* microRNA-mediated repression of protein synthesis. Using this system, we found that a target mRNA was deadenylated when it was translationally repressed. The experimental data strongly suggested that the deadenylation was a cause, but not a result, of translational repression. In this chapter, we describe our cell-free system and discuss the significance of microRNA-mediated mRNA deadenylation in the repression of protein synthesis.

M. Wakiyama and S. Yokoyama (✉)
RIKEN, Systems and Structural Biology Center, 1-7-22 Suehiro-cho, Tsurumi,
Yokohama 230-0045, Japan
e-mail: waki@gsc.riken.jp

S. Yokoyama (✉)
Graduate School of Science, The University of Tokyo, 7-3-1 Hongo, Bunkyo,
Tokyo, 113-0033, Japan
e-mail: yokoyama@biochem.s.u-tokyo.ac.jp

R.E. Rhoads (ed.), *miRNA Regulation of the Translational Machinery,*
Progress in Molecular and Subcellular Biology, 50,
DOI 10.1007/978-3-642-03103-8_6, © Springer-Verlag Berlin Heidelberg 2010

6.1 Introduction

Cell-free systems have been used for decades for studying protein synthesis. They have made a significant contribution to the decoding of genetic codes. Experiments with cell-free systems have provided valuable information about the mechanisms of post-transcriptional regulation, including mRNA stability and translation. Cell-free systems have many advantages over in vivo systems. By using cytoplasmic extracts in combination with in vitro transcripts, we can focus on the post-transcriptional events of gene expression. In the cell-free systems, it is very easy to modify various reaction conditions, to compare the translation efficiency and stability of different mRNA constructs, and to monitor the time courses of reactions.

Various cell-free systems using different biological sources and protocols have been developed. The most widely used eukaryotic system is rabbit reticulocyte lysate (RRL) (Jackson and Hunt 1983). The RRL is prepared by lysing red blood cells obtained from rabbits recovering from experimentally induced anemia. This system has great advantages in its simplicity of preparation and relatively high protein synthesis activity. Thus, it became the most popular system used for studying translation mechanisms. Many of the canonical translation factors were purified and characterized using the RRL system.

An appropriate cell-free system will make a significant contribution to the biochemical dissection of miRNA mechanisms. Thus, we sought to establish a cell-free system that recapitulates many of the phenomena of miRNA-mediated repression of protein synthesis. There are two possible strategies to reconstitute miRNA function. One is to develop a cell-free system based on endogenous miRNA/miRISC activity. This strategy requires materials containing validated miRNA, which is relatively abundant. Thermann and Hentze chose the regulation of *Drosophila melanogaster* reaper mRNA by miR2 in *Drosophila* embryos (Thermann and Hentze 2007). Sonenberg and coworkers utilized extracts from Krebs-2 ascites cells, which express *let-7* miRNA, the best characterized miRNA so far (Mathonnet et al. 2007). Another option for establishing a cell-free system is to use chemically synthesized short RNA fragments that mimic miRNA or preform miRNA. It is advantageous to use exogenous miRNA, since we can analyze the formation of miRISCs and translation of the reporter mRNA both in the presence and absence of the miRNA. We have adopted this strategy in our experiments. In addition, Novina and coworkers performed experiments using the RRL system with the CXCR4 siRNA duplex as an miRNA mimic (Wang et al. 2006).

In this chapter, we describe the recapitulation of *let-7* miRNA-mediated repression of protein synthesis in a cell-free system, which was established with extracts prepared from HEK293F cells overexpressing miRNA pathway components (Wakiyama et al. 2007).

6.2 Cell-Free Systems that Recapitulate RNA Silencing

A cell-free system for studying the mechanism of RNA silencing was first described by Tuschl and coworkers (Tuschl et al. 1999). This represented the pioneering work of the biochemical dissection of RNAi. In our attempts to develop cell-free assays to analyze miRNA functions in translation, we performed many preliminary experiments using *Drosophila* Schneider 2 (S2) cells as the extract source. We succeeded in recapitulating RNAi with the S2 extracts (Wakiyama et al. 2006). In this system, the GL3 firefly luciferase (FLuc) mRNA was specifically targeted with the GL3 siRNA, but not with the GL2 siRNA, which differs by only three nucleotides from the GL3 siRNA. Although the RNAi assays were successful, we were not able to see the effects of the miRNAs in this cell-free system. We used an siRNA-like 21 nucleotide duplex, in which one strand contained the *Drosophila let-7* miRNA sequence. Although it worked as an siRNA targeting mRNA containing the perfect complementary sequence to *let-7*, it did not repress the translation of mRNA containing multiple *let-7* target sites, which was half complementary to *let-7* (unpublished data).

Gregory and coworkers showed that pre-miRNA processing and assembly of RISC, an RNA-induced silencing complex, are functionally coupled (Gregory et al. 2005). Thus, we used the preform of *Drosophila let-7* miRNA, *Dlet-7* pre-miRNA, which is 60 nucleotides long. First, we looked at whether the *let-7* pre-miRNAs were processed into the mature *let-7* miRNAs in our S2 cell extracts, and found that the processing efficiency was very poor. Indeed, no repression was observed for the mRNA containing *let-7* target sites. At this point, we assumed that something required for the efficient processing of pre-miRNA was missing in our S2 extracts.

We decided to try material from mammalian cells, instead of the *Drosophila* S2 cells. We analyzed HEK293F cells, which are available from Invitrogen. HEK293F cells, a subtype of HEK293, derived from human embryonic kidney, can grow in suspension in serum-free medium, FreeStyle293 Expression medium (Invitrogen). A great advantage of HEK293F cells is that external DNA can be introduced and expressed with high efficiency by transient transfection. We detected expression from more than 50% of the cells, by monitoring with EGFP. Therefore, we assumed that we could easily prepare lysates from cells overexpressing a specific component of the miRNA pathway.

6.3 Cell Extract Preparation Methodology

To establish a cell-free system, the cell extract preparation method should be extensively investigated. We employed the nitrogen cavitation method (Wakiyama et al. 2006, 2007). Nitrogen cavitation has been used with various cells and tissues to prepare biomacromolecules, such as polyribosomes and microsomes (Dowben et al. 1968; Short et al. 1972). This technique involves the equilibration of nitrogen gas with a cell suspension under high pressure, followed by sudden decompression. The use of nitrogen gas prevents thermal deactivation and oxidization of biochemical macromolecules. Disruption of nuclei is avoided by choosing appropriate conditions for the lysis solution and the pressure of the nitrogen gas.

6.4 Processing of Pre-miRNA

We first examined the processing of *Drosophila let-7* (*Dlet-7*) pre-miRNA in HEK293F extracts. Similar to our observations with the S2 cell extracts, only a fraction of the *let-7* pre-miRNA was converted to its mature form. We then tested different factors to add to the extracts for efficient processing. We combined extracts expressing Dicer, TRBP2, Argonaute2, and GW182 in several combinations, and performed processing assays of the *let-7* pre-miRNA (Fig. 6.1). Interestingly, an extra amount of Dicer, a processing enzyme for the miRNA, did not affect processing. However, Argonaute2 (or Argonaute1) remarkably improved the processing efficiency. GW182, which interacts with the Argonaute protein, did not affect the

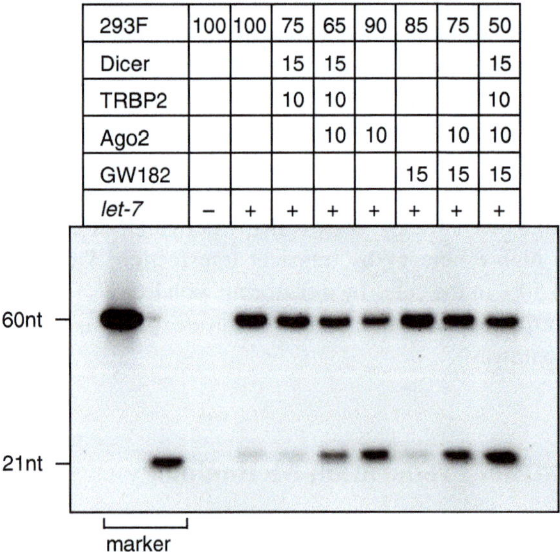

Fig. 6.1 Processing assay of *let-7* pre-miRNA. Chemically synthesized *let-7* pre-miRNA was incubated with extracts prepared from HEK293F cells, or the mixed extracts composed of the extracts prepared from HEK293F cells overexpressing Dicer, TRBP2, Ago2 or GW182, at the ratios indicated above the figure. The RNA was then extracted, and the *let-7* was detected by northern blotting. Markers are a 60 nucleotide (60nt) *let-7* pre-miRNA and a 21nt *let-7* siRNA. Reprinted from Wakiyama et al. 2007 with permission from Cold Spring Harbor Laboratory Press

processing efficiency. We assumed that an extra amount of the Argonaute proteins might be required for the efficient incorporation of externally added miRNAs into miRISCs.

6.5 In Vitro Translation Assay

To assay the effects of *let-7*, we constructed FLuc reporter mRNAs containing multiple *let-7* target sequences, derived from the *Drosophila lin-41* ortholog, in the 3′-UTR (Fig. 6.2) (Pasquinelli et al. 2000). As a control, we also made a mutant construct, in which the seeding region of the *let-7* target sequence, UACCUC, was converted to AUGGAG (Fig. 6.2). In both the in vivo and in vitro experimental systems used thus far, multiple target sites in the 3′-UTR of the reporter mRNA enhanced miRNA-mediated gene silencing. Indeed, higher numbers of *let-7* target sites in the 3′-UTR of our FLuc reporter mRNA enhanced the inhibition of FLuc synthesis (unpublished data). This fact implies that the molar ratio between the miRISC and the target sites may be critical for the miRNA-mediated repression of protein synthesis.

Figure 6.3 shows the translation of capped and polyadenylated FLuc-6xT mRNA, which contains six *let-7* target sequences in the 3′-UTR, in the absence (−) and presence (+) of *let-7*. The *Renilla* luciferase mRNA was simultaneously translated, to normalize the FLuc activities in different experiments. In the presence of *let-7*, the synthesis of FLuc was reduced to ~70% in the HEK293F cell extracts, while it was about ~50% in the cell extracts composed of HEK293F and HEK293F overexpressing Argonaute2 (Ago2). This can be explained by the finding that overexpressed Ago2 stimulates pre-miRNA processing and probably the formation of miRISCs containing *let-7* miRNA. Interestingly, the addition of an extract, prepared from HEK293 cells overexpressing GW182, significantly increased *let-7*-mediated repression to ~70%. GW182 is a key component of P/GW-body, a discrete cytoplasmic compartment where translational repression and mRNA turnover may occur. We also showed that FLAG-tagged Ago2 and FLAG-tagged GW182 were specifically recruited to the reporter mRNA in the presence of *let-7* miRNA, by using a biotinylated capped and polyadenylated FLuc-6xT mRNA (Fig. 6.4). These data indicated that GW182 is involved in the miRNA-mediated

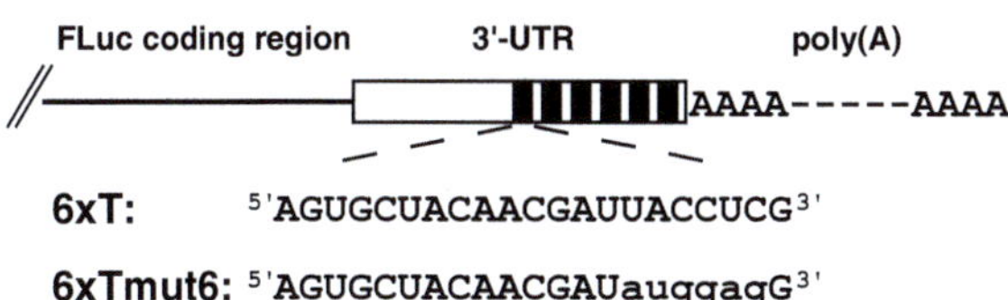

Fig. 6.2 Schematic representations of the FLuc mRNAs FLuc-6xT and FLuc-6xTmut6. Reprinted from Wakiyama et al. 2007 with permission from Cold Spring Harbor Laboratory Press

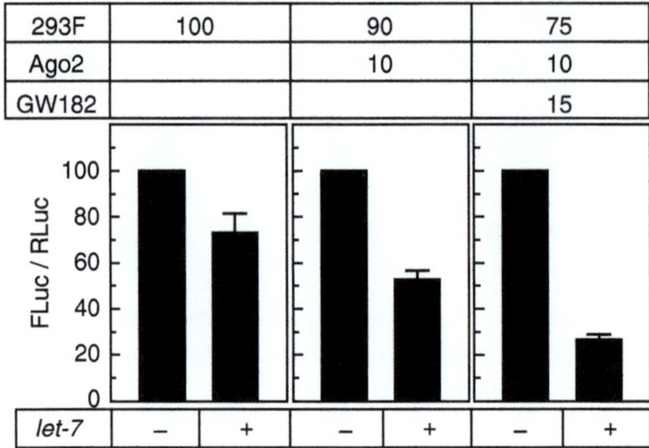

Fig. 6.3 In vitro translation of capped and polyadenylated FLuc-6xT mRNA in the absence (−) and presence (+) of *let-7*. Cell extracts prepared from control HEK293F cells, and HEK293F cells overexpressing FLAG-Ago2 and FLAG-GW182 were used in the combinations indicated above the figure. The FLuc and RLuc activities were measured, and the FLuc to RLuc activity ratio in the reaction without *let-7* was set at 100. The data shown constitute an average of at least three independent experiments with standard deviations. Reprinted from Wakiyama et al. 2007 with permission from Cold Spring Harbor Laboratory Press

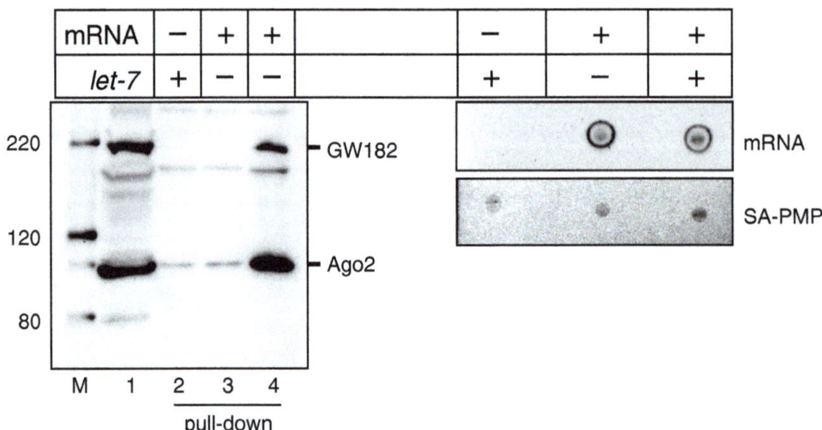

Fig. 6.4 Ago2 and GW182 are recruited to the reporter mRNA in a *let-7*-dependent manner. Biotinylated FLuc-6xT mRNAs (capped and polyadenylated) were incubated with the mixed extract (Ago2 + GW182, see lane 1) in the absence (−) and presence (+) of *let-7*. The mRNAs were captured with Streptavidin Paramagnetic Particles (SA-PMPs, Promega). *Left panel*: The proteins bound to the mRNAs were detected using an anti-FLAG M2 antibody (lane 1: Ago2 + GW182 extract; lanes 2–4: pull-down). The sizes (kDa) of the molecular markers (M) are indicated on the left. *Right panel*: The mRNAs captured with SA-PMPs were detected by dot hybridization. Reprinted from Wakiyama et al. 2007 with permission from Cold Spring Harbor Laboratory Press

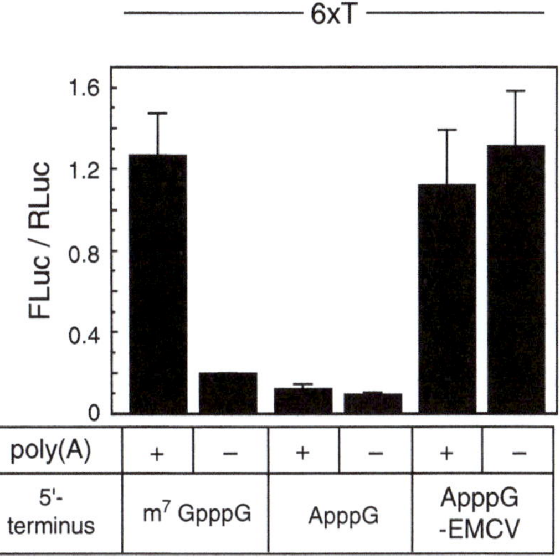

Fig. 6.5 Comparison of translation efficiencies of different constructs of FLuc-6xT reporter mRNAs in the cell-free system. Reprinted and modified from Wakiyama et al. 2007 with permission from Cold Spring Harbor Laboratory Press

repression of protein synthesis. Recently, we found that GW182 contains multiple binding sites for Argonaute. Furthermore, we demonstrated that the interaction between GW182 and Argonaute is critical for the *let-7* miRNA-mediated repression in the cell-free system. While *let-7* repressed the translation of capped and polyadenylated FLuc-6xT mRNA, it did not affect the translation of capped and polyadenylated FLuc-6xTmut6, in which the seeding region was mutated. In addition, *let-7*-mediated repression was relieved by the addition of anti-*let-7* RNA (data not shown). These results indicated that the *let-7*-mediated repression of protein synthesis in our cell-free system is sequence-specific.

One of the most controversial topics of the miRNA mechanism is whether miRNA inhibits the initiation of translation. Therefore, we examined the effects of *let-7* on several mRNA constructs. We prepared FLuc-6xT mRNAs with a nonphysiological A(5′)ppp(5)G cap (ApppG) and FLuc-6xT mRNAs with ApppG-capped EMCV IRES. The ApppG cap is not recognized by the translation initiation factor eIF4E, and thus, it does not function in canonical translation initiation. The ApppG-capped mRNA was appreciably translated in our cell-free system, although its translation efficiency was lower than that of the physiological m⁷GpppG-capped mRNA (Fig. 6.5). The mRNA with the IRES, internal ribosome entry site, derived from EMCV is translated by an eIF4E-independent mechanism. In our system, the translation efficiency of the mRNA with EMCV IRES was comparable to that of the m⁷GpppG-capped and polyadenylated mRNA (Fig. 6.5). As shown in Fig. 6.6, it is obvious that the *let-7*-mediated repression of protein synthesis in our cell-free system requires both the 5′-cap structure and 3′-poly(A) tail. The FLuc synthesis from the ApppG-EMCV IRES-FLuc-6xT mRNAs was slightly repressed by *let-7*, especially when they were polyadenylated.

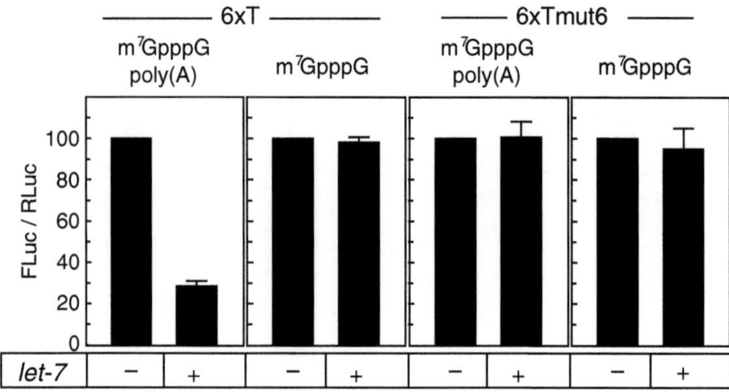

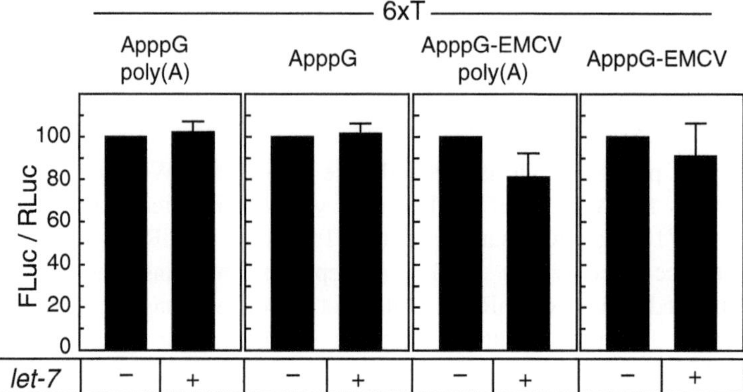

Fig. 6.6 The m⁷GpppG cap and poly(A) tail are important for the *let-7*-mediated translational repression. Reprinted and modified from Wakiyama et al. 2007 with permission from Cold Spring Harbor Laboratory Press

6.6 *Let-7*-Mediated mRNA Deadenylation

In order to examine whether *let-7* miRNA changes the abundance of target mRNAs, we performed northern blotting (Fig. 6.7). The capped and polyadenylated FLuc-6xT mRNAs, but not the capped and polyadenylated FLuc-6xTmut6 mRNAs, were obviously shorter, and nearly the same length as the nonadenylated mRNAs, after a 60-min incubation in the presence of *let-7*. This result raises the possibility that the target mRNA was deadenylated in a *let-7*-dependent manner. Therefore, we analyzed the poly(A) status of these mRNAs by an RNaseH cleavage assay (Fig. 6.7). In this assay, RNAs were extracted from cell extracts at the end of the translation reaction, hybridized with oligo(dT), and then treated with RNaseH. After the RNaseH treatment, the bands of the FLuc-6xT mRNAs shifted to almost the same position as the nonadenylated mRNAs under all conditions tested.

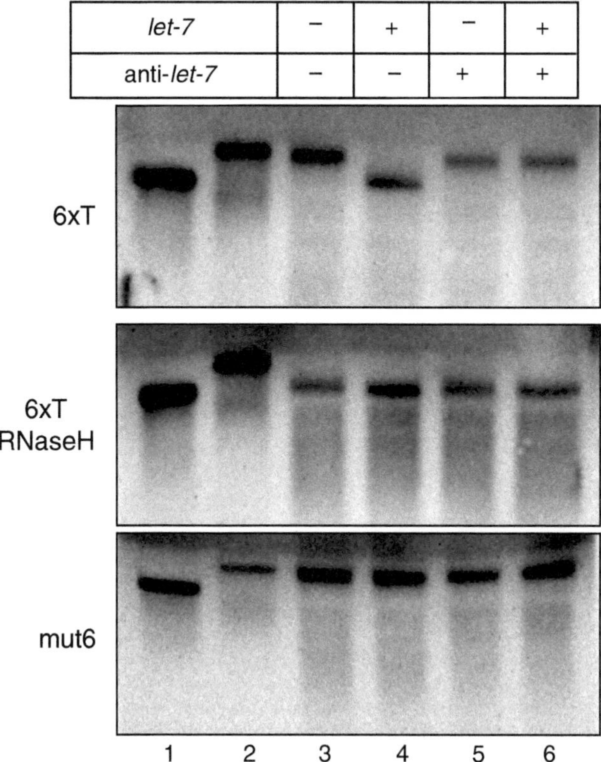

Fig. 6.7 *Let-7* miRNAs direct deadenylation of the target mRNAs. Analyses of capped and polyadenylated FLuc-6xT (6xT) and FLuc-6xTmut6 (mut6) mRNAs by northern blotting. As molecular markers, nonadenylated and polyadenylated transcripts (capped) were loaded in lanes 1 and 2, respectively. After 60 min of translation in the absence (−) and presence (+) of *let-7* and anti-*let-7*, the RNAs were extracted from the reaction mixture, and similar amounts were loaded in lanes 3–6. In the *middle panel*, the extracted RNAs, as in the *upper panel* (6xT), were annealed with oligo(dT) and subjected to RNaseH digestion. Reprinted from Wakiyama et al. 2007 with permission from Cold Spring Harbor Laboratory Press

These results support the idea that the FLuc-6xT mRNAs are deadenylated in a *let-7*-dependent manner. Further experiments revealed that the deadenylation occurred regardless of the presence or absence of the 5′-cap structure. Moreover, the *let-7*-dependent deadenylation still occurred in the presence of cycloheximide, an inhibitor of translation, and thus it is independent of translation.

To further analyze the deadenylation and translation, we performed parallel time-course analyses of the luciferase expression and mRNA deadenylation, with the m^{7}GpppG-capped and polyadenylated FLuc-6xTmRNA (Fig. 6.8). The data indicated a strong correlation between mRNA deadenylation and repression of FLuc synthesis.

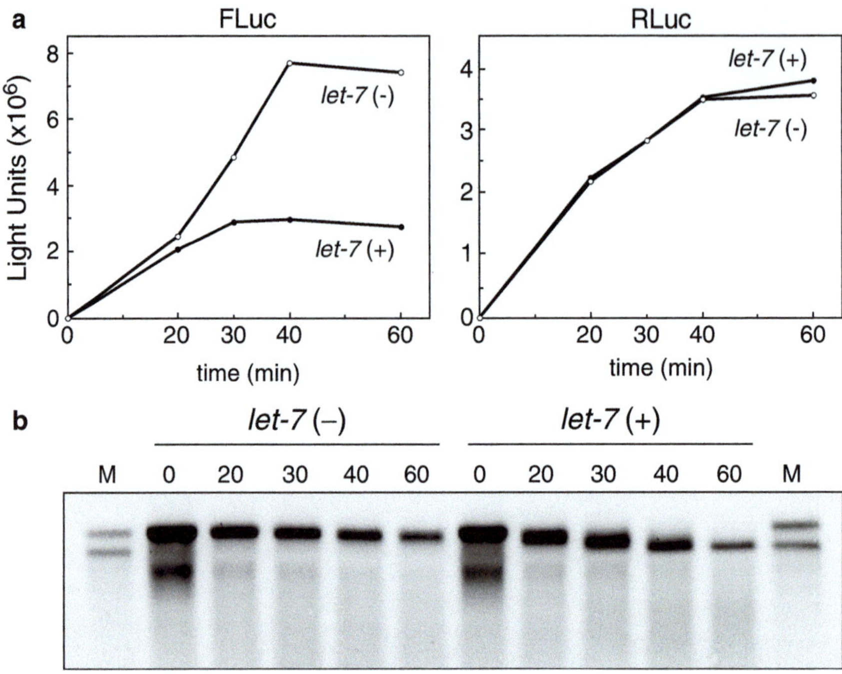

Fig. 6.8 Deadenylation of the reporter mRNA and translational repression. (**a**) Time-course analysis of luciferase synthesis. (**b**) RNAs were extracted from the aliquots taken at each time point, and were subjected to a northern hybridization analysis using the FLuc probe. Reprinted from Wakiyama et al. 2007 with permission from Cold Spring Harbor Laboratory Press

6.7   Poly(A) tail and miRNA-Mediated Repression

The involvement of the poly(A) tail in miRNA-mediated repression has been reported by several groups. Humphreys and coworkers demonstrated that an m⁷G-cap was critical for repression, but a poly(A) tail was also important for efficient repression (Humphreys et al. 2005). Wang and coworkers reported the first successful in vitro recapitulation of miRNA-mediated repression (Wang et al. 2006). They used a RRL with an artificial miRNA mimic and a luciferase mRNA with multiple target sites. The miRNA was preannealed to the target mRNA before it was added to the RRL. A m⁷GpppG-cap and a poly(A) tail were important for the repression in this system. Moreover, uncapped mRNAs were also repressed if they had a long poly(A) tail, extended to about 2,000 residues. MiRNA-mediated deadenylation of target mRNA has been reported in several in vivo systems, including *Drosophila* S2 cells, HEK293T cells and zebrafish (Behm-Ansmant et al. 2006; Giraldez et al. 2006; Wu et al. 2006). Collectively, deadenylation is one of the major effects of miRNA regulation.

6.8 Poly(A) tail and Post-Transcriptional Regulation

The concept that the 5′ and 3′ ends of eukaryotic mRNA communicate in the process of post-transcriptional regulation is supported by various biochemical experiments. Circular shaped polysomes have been observed by electron microscopy and atomic force microscopy (Yoshida et al. 1997). The 5′ cap structure and the poly(A) tail, the key features of most of eukaryotic cellular mRNAs, are involved in the communication of the two ends of eukaryotic mRNAs. The cap structure is recognized by the eukaryotic translation initiation factor eIF4E, a component of the eIF4F complex (Gingras et al. 1999). Another eIF4F component, eIF4G, binds a poly(A)-binding protein, PABP (Sachs 2000). The interaction between eIF4G and PABP is critical for efficient translation. PABP functions as a translational initiation factor and enhances the formation of the initiation complex (Kahvejian et al. 2005). In *Xenopus*, interference with the interaction of PABP and eIF4G results in the inhibition of poly(A)-dependent translation and oocyte maturation (Wakiyama et al. 2000). Many maternal mRNAs are regulated by the length of the poly(A) tail. Typically, the translation of these mRNAs is inhibited when their poly(A) tails are shortened, and is activated when their poly(A) tails are elongated. These processes are controlled by proteins that bind to the 3′-UTR of the target mRNAs (Wickens et al. 2002). CPE, a cytoplasmic polyadenylation element, and the CPE-binding protein, CPEB, are involved in this regulation (Mendez and Richter 2001). CPEB also binds Maskin, a protein that interacts with eIF4E (Richter and Sonenberg 2005). This interaction disrupts eIF4E–eIF4G formation and results in the inhibition of translation. An analogous inhibitory regulation mechanism has been discovered in *Drosophila*. The eIF4E-binding protein, Cup, is a Maskin-like protein that controls germ-cell formation and axis specification (Richter and Sonenberg 2005). Cup interacts with Bruno, which binds to the Bruno response element in the 3′-untranslated region (3′-UTR) of the *oskar* mRNA. The Cup–eIF4E interaction prevents the assembly of eIF4E and eIF4G, which in turn inhibits the translation of the *oskar* mRNA. The cross talk between the 5′ and 3′ ends of the mRNA is critical not only for translation but also for mRNA deadenylation. A poly(A)-specific ribonuclease, PARN, catalyzes the deadenylation of maternal mRNAs during *Xenopus* oocyte maturation. PARN contains the cap-binding pocket, and its activity is stimulated by the cap structure (Dehlin et al. 2000).

6.9 Concluding Remarks

In our cell-free system, repression occurred only for a m^7GpppG-capped and polyadenylated mRNA, while deadenylation occurred regardless of the 5′ structure of the mRNA or active translation. The mRNA was not degraded when the translation stopped, which means that deadenylation did not lead to rapid degradation of FLuc mRNAs. Therefore, we propose that miRNA-mediated mRNA deadenylation abolishes the cap-poly(A) synergy and represses translation (Fig. 6.9).

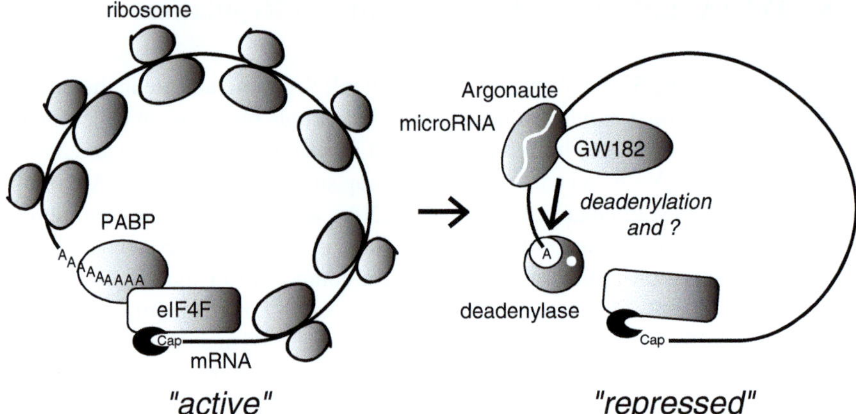

Fig. 6.9 Model of miRNA-mediated mRNA deadenylation and repression of protein synthesis

However, this model has been challenged by results, obtained in other systems, which showed that nonadenylated mRNAs are silenced. Wu and coworkers reported that a *let-7* target mRNA with a histone 3′ stem-loop in place of the poly(A) tail was repressed (Wu et al. 2006). Recently, Eulalio and coworkers demonstrated that miRNAs also silence mRNAs with a 3′ end generated by ribozyme cleavage (Eulalio et al. 2009). How can these discrepancies be explained? Tomari and coworkers recently proposed that *Drosophila* Ago1 and Ago2 may inhibit protein synthesis by a different mechanism (by personal communication). It is possible that miRNAs repress protein synthesis by multiple mechanisms, including deadenylation-directed translational inhibition.

Acknowledgments This work was supported by the RIKEN Structural Genomics/Proteomics Initiative (RSGI), and by the National Project on Protein Structural and Functional Analysis, Ministry of Education, Culture, Sports, Science and Technology of Japan.

References

Behm-Ansmant I, Rehwinkel J, Doerks T, Stark A, Bork P, Izaurralde E (2006) mRNA degradation by miRNAs and GW182 requires both CCR4:NOT deadenylase and DCP1:DCP2 decapping complexes. Genes Dev 20:1885–1898

Dehlin E, Wormington M, Korner CG, Wahle E (2000) Cap-dependent deadenylation of mRNA. EMBO J 19:1079–1086

Dowben RM, Gaffey A, Lynch PM (1968) Isolation of liver and muscle polyribosomes in high yield after cell disruption by nitrogen cavitation. FEBS Lett 2:1–3

Eulalio A, Huntzinger E, Nishihara T, Rehwinkel J, Fauser M, Izaurralde E (2009) Deadenylation is a widespread effect of miRNA regulation. RNA 15:21–32

Gingras AC, Raught B, Sonenberg N (1999) eIF4 initiation factors: effectors of mRNA recruitment to ribosomes and regulators of translation. Annu Rev Biochem 68:913–963

Giraldez AJ, Mishima Y, Rihel J, Grocock RJ, Van Dongen S, Inoue K, Enright AJ, Schier AF (2006) Zebrafish MiR-430 promotes deadenylation and clearance of maternal mRNAs. Science 312:75–79

Gregory RI, Chendrimada TP, Cooch N, Shiekhattar R (2005) Human RISC couples microRNA biogenesis and posttranscriptional gene silencing. Cell 123:631–640

Humphreys DT, Westman BJ, Martin DI, Preiss T (2005) MicroRNAs control translation initiation by inhibiting eukaryotic initiation factor 4E/cap and poly(A) tail function. Proc Natl Acad Sci U S A 102:16961–16966

Jackson RJ, Hunt T (1983) Preparation and use of nuclease-treated rabbit reticulocyte lysates for the translation of eukaryotic messenger RNA. Methods Enzymol 96:50–74

Kahvejian A, Svitkin YV, Sukarieh R, M'Boutchou MN, Sonenberg N (2005) Mammalian poly(A)-binding protein is a eukaryotic translation initiation factor, which acts via multiple mechanisms. Genes Dev 19:104–113

Mathonnet G, Fabian MR, Svitkin YV, Parsyan A, Huck L, Murata T, Biffo S, Merrick WC, Darzynkiewicz E, Pillai RS, Filipowicz W, Duchaine TF, Sonenberg N (2007) MicroRNA inhibition of translation initiation in vitro by targeting the cap-binding complex eIF4F. Science 317:1764–1767

Mendez R, Richter JD (2001) Translational control by CPEB: a means to the end. Nat Rev Mol Cell Biol 2:521–529

Pasquinelli AE, Reinhart BJ, Slack F, Martindale MQ, Kuroda MI, Maller B, Hayward DC, Ball EE, Degnan B, Muller P, Spring J, Srinivasan A, Fishman M, Finnerty J, Corbo J, Levine M, Leahy P, Davidson E, Ruvkun G (2000) Conservation of the sequence and temporal expression of let-7 heterochronic regulatory RNA. Nature 408:86–89

Richter JD, Sonenberg N (2005) Regulation of cap-dependent translation by eIF4E inhibitory proteins. Nature 433:477–480

Sachs A (2000) Physical and functional interactions between the mRNA cap structure and the poly(A) tail. In Sonenberg N, Hershey JWB, Mathews MB (eds) Translational control of gene expression. Cold Spring Harbor Laboratory Press, Cold Spring Harbor, New York, pp 447–465

Short CR, Maines MD, Davis LE (1972) Preparation of hepatic microsomal fraction for drug metabolism studies by rapid decompression homogenization. Proc Soc Exp Biol Med 140:58–65

Thermann R, Hentze MW (2007) Drosophila miR2 induces pseudo-polysomes and inhibits translation initiation. Nature 447:875–878

Tuschl T, Zamore PD, Lehmann R, Bartel DP, Sharp PA (1999) Targeted mRNA degradation by double-stranded RNA in vitro. Genes Dev 13:3191–3197

Wakiyama M, Imataka H, Sonenberg N (2000) Interaction of eIF4G with poly(A)-binding protein stimulates translation and is critical for Xenopus oocyte maturation. Curr Biol 10:1147–1150

Wakiyama M, Kaitsu Y, Yokoyama S (2006) Cell-free translation system from Drosophila S2 cells that recapitulates RNAi. Biochem Biophys Res Commun 343:1067–1071

Wakiyama M, Takimoto K, Ohara O, Yokoyama S (2007) Let-7 microRNA-mediated mRNA deadenylation and translational repression in a mammalian cell-free system. Genes Dev 21: 1857–1862

Wang B, Love TM, Call ME, Doench JG, Novina CD (2006) Recapitulation of short RNA-directed translational gene silencing in vitro. Mol Cell 22:553–560

Wickens M, Bernstein DS, Kimble J, Parker R (2002) A PUF family portrait: 3'UTR regulation as a way of life. Trends Genet 18:150–157

Wu L, Fan J, Belasco JG (2006) MicroRNAs direct rapid deadenylation of mRNA. Proc Natl Acad Sci U S A 103:4034–4039

Yoshida T, Wakiyama M, Yazaki K, Miura K (1997) Transmission electron and atomic force microscopic observation of polysomes on carbon-coated grids prepared by surface spreading. J Electron Microsc 46:503–506

Chapter 7
miRNA Effects on mRNA Closed-Loop Formation During Translation Initiation

Traude H. Beilharz, David T. Humphreys, and Thomas Preiss

Contents

Abstract A flurry of recent studies, carried out primarily in transfected cells or in vitro translation systems, have attempted to reveal the molecular means by which animal microRNAs (miRNAs) attenuate mRNA translation. Despite these intense efforts it has not yet been possible to derive a consensus model for such a mechanism. Here we summarise our own experimental contributions to this topic, which led us to propose that miRNAs control early translation initiation by affecting eukaryotic initiation factor 4E/cap structure and poly(A) tail function, and place them in a current context of this rapidly moving and challenging field.

7.1   Introduction

Any contemporary view of eukaryotic gene expression has to allow for regulatory roles of the range of non-coding RNAs known to originate from most regions of complex genomes (Amaral et al. 2008). MicroRNAs (miRNAs) comprise a tiny but

T.H. Beilharz, D.T. Humphreys, and T. Preiss (✉)
Molecular Genetics Division, Victor Chang Cardiac Research Institute (VCCRI), Darlinghurst, Sydney, NSW 2010, Australia
e-mail: t.preiss@victorchang.edu.au

T.H. Beilharz and T. Preiss
School of Biotechnology and Biomolecular Sciences and St Vincent's Clinical School, University of New South Wales, Sydney, NSW 2052, Australia

R.E. Rhoads (ed.), *miRNA Regulation of the Translational Machinery*,
Progress in Molecular and Subcellular Biology, Vol. 50,
DOI 10.1007/978-3-642-03103-8_7, © Springer-Verlag Berlin Heidelberg 2010

well-recognised portion of this potentially vast regulatory repertoire of cells. It is now firmly established that the ~22-nucleotide-long miRNAs have important physiological and pathophysiological roles (Carthew 2006; Kloosterman and Plasterk 2006; Chang and Mendell 2007), and much progress has been made to understand their biogenesis (Faller and Guo 2008). Our interest in miRNAs began when they emerged as a new and populous class of translational regulators with pervasive effects on the expression of cellular transcriptomes (Bartel and Chen 2004; Friedman et al. 2008). miRNAs interact with argonaute proteins (Peters and Meister 2007) and guide RNA-induced silencing complexes (RISC or miRNP) to target mRNAs (Kloosterman and Plasterk 2006; Chang and Mendell 2007). Binding of animal miRNAs to imperfectly matching sequences, usually in the 3′ untranslated region (UTR) of their target mRNAs, inhibits accumulation of the encoded proteins by reducing mRNA stability and/or translation (Valencia-Sanchez et al. 2006; Eulalio et al. 2008a; Filipowicz et al. 2008; Wu and Belasco 2008). Investigations with strongly destabilised targets revealed that miRNA can trigger mRNA deadenylation and exonucleolytic decay (Valencia-Sanchez et al. 2006; Eulalio et al. 2008a), in part explaining the emerging links of miRNA function with the constituents and function of cytoplasmic processing bodies, known sites of mRNA storage and decay (Eulalio et al. 2007).

We, and others, have further studied the effects of miRNAs on translation under conditions of little mRNA destabilisation (Humphreys et al. 2005; Pillai et al. 2005;

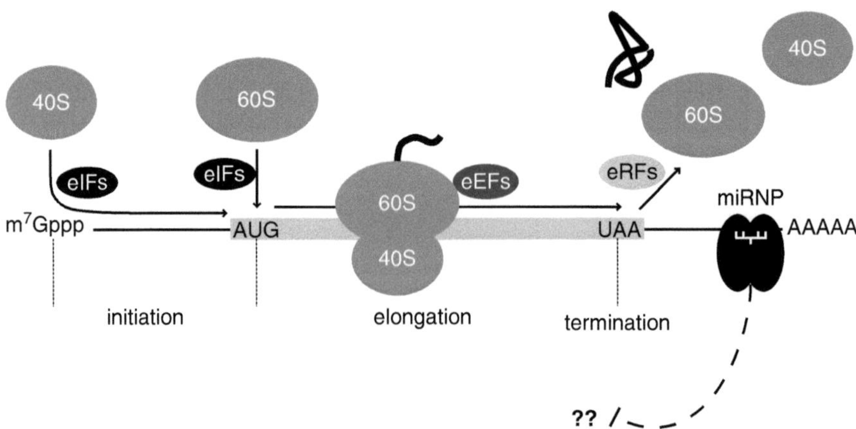

Fig. 7.1 Schematic view of the phases of translation on a typical cellular mRNA that may be targeted by an miRNA. During initiation, the 40S ribosomal subunit is recruited to the 5′ end of the mRNA and scans the 5′ UTR to locate the initiator codon, where it is joined by the 60S subunit to form the 80S ribosome. All steps of initiation require a multitude of accessory factors termed eIFs (eukaryotic initiation factor). During elongation, 80S ribosomes traverse the coding region and synthesise the encoded polypeptide. This requires a separate set of factors termed eEFs (eukaryotic elongation factor). The stop codon signals termination, the dissociation of ribosomal subunits and the finished polypeptide from the mRNA. Eukaryotic release factors (eRFs) participate in this process. *A dotted line* and *question marks* are shown to illustrate the incompletely understood mechanisms of translational repression by a miRNP bound to the mRNA 3′ UTR

Table 7.1 Some recent reports on the mechanism of translational repression by miRNA

Study	Step(s)	mRNA feature(s)	Factor(s)/comment	miRNA/system
Pillai et al. (2005)	Initiation	Cap	eIF4E	Let-7/ transfected HeLa cells
Humphreys et al. (2005)	Initiation	Cap and tail	eIF4E	miCXCR4/ transfected HeLa cells
Petersen et al. (2006)	Elongation/Ribosome drop-off	–	–	miCXCR4/ transfected 293T cells
Wang et al. (2006)	Initiation	Cap and tail	–	miCXCR4/ in vitro (RRL)
Nottrott et al. (2006)	Proteolysis of nascent peptide	–	–	let-7/ transfected HeLa cells
Maroney et al. (2006)	Elongation	–	–	Endogenous miRNA and targets in HeLa cells
Thermann and Hentze (2007)	Initiation	Cap	–	miR2 in fly embryo extracts
Lytle et al. (2007)	Post-initiation	–	Dependence on transfection method	Let-7/ transfected HeLa cells
Wakiyama et al. (2007)	Initiation	Cap and tail	–	Let-7/ in vitro (modified HEK293F cell extracts)
Mathonnet et al. (2007)	Initiation	Cap	eIF4F	Let-7/ in vitro (Krebs -2 ascites cell extracts)
Kong et al. (2008)	Initiation or post-initiation	–	Dependence on promoter identity/ nuclear history	Let-7 and miR-34/ transfected HeLa cells

Maroney et al. 2006; Nottrott et al. 2006; Petersen et al. 2006; Wang et al. 2006; Lytle et al. 2007; Mathonnet et al. 2007; Thermann and Hentze 2007; Wakiyama et al. 2007; Kong et al. 2008). The central question in each of these studies has been to understand what phase of the translation process is primarily affected and what processes within that phase are molecular targets of the repressive mechanism(s) emanating from the miRNP bound to the mRNA (illustrated in Fig. 7.1). Despite the use of very similar tools and experimental approaches, these studies have led to dramatically different conclusions as summarised in Table 7.1. The findings and models derived from several of these endeavours in other laboratories are detailed in other chapters of this volume. The issue thus continues to be actively debated and the salient controversies and arguments are also well covered in a number of excellent recent review articles (Standart and Jackson 2007; Eulalio et al. 2008a; Filipowicz et al. 2008; Wu and Belasco 2008).

7.2 Analysing Translational Repression by an miRNA

The mRNA translation process comprises of initiation, elongation and termination phases, each requiring a unique set of auxiliary factors in addition to ribosomes (Fig. 7.1). Eukaryotic translation initiation depends on multiple eukaryotic initiation factors (eIF) for an elaborate, step-wise recruitment of ribosomes to the start of the mRNA coding region (Preiss and Hentze 2003; Sonenberg and Dever 2003; Kapp and Lorsch 2004). Most known mechanisms for the control of eukaryotic translation affect the initiation phase, although regulation of elongation, termination or nascent polypeptide stability have also been reported (Preiss and Hentze 2003; Sonenberg and Dever 2003; Gebauer and Hentze 2004). Over many years, a range of experimental tools have been developed to investigate translational control, and these tools are increasingly applied to study the role of miRNAs. For instance, the pioneers of miRNA research in *Caenorhabditis elegans* studied the regulation of *lin-14* and *lin-28* mRNAs using sucrose density gradient ultracentrifugation. These studies found that both mRNAs co-sedimented with polyribosomes, even when isolated from worm larval stages where their expression is repressed by miRNAs, leading to the first suggestion of a model for miRNA action, namely that miRNAs affect the post-initiation phase of translation (Olsen and Ambros 1999; Seggerson et al. 2002; Ambros 2004).

7.2.1 Use of Viral Internal Ribosome Entry Sites
to Study the miRNA Mechanism

In our own work, we decided to employ viral internal ribosome entry sites (IRES) as principal tools to investigate which sub-step of translation is targeted by an miRNA. This followed the previously established premise that if a control mechanism under study targets a component of the canonical translation machinery not required for IRES-mediated translation, then translation driven by the IRES should no longer be subject to regulation (Ostareck et al. 2001; Poyry et al. 2004; Jackson 2005). The IRES elements listed in Table 7.2 are suitable for this type of analysis, as careful biochemical and structural analyses have defined different initiation factor requirements for translation in each case (Borman and Kean 1997; Hellen and Sarnow 2001; Poyry et al. 2004; Fraser and Doudna 2006).

To implement this approach, we adopted a well-studied synthetic miRNA/target reporter pairing (Doench et al. 2003; Doench and Sharp 2004). In the original publications, the synthetic RNA duplex (termed miCXCR4 as it was originally designed as an siRNA against CXCR4 mRNA) was co-transfected together with a plasmid (pRL-TK-4 sites) that expresses a *Renilla*-luciferase mRNA containing four imperfectly matching binding sites for miCXCR4 in its 3′ UTR (R-luc-4 sites). Deployed in this way, miCXCR4 acted like a genuine miRNA by inducing robust and specific repression of R-luc protein expression with little destabilisation of the plasmid-derived target mRNA (Doench et al. 2003; Doench and Sharp 2004; Humphreys et al. 2007). To facilitate the engineering of predictable changes in the mode of translation occurring on the reporter mRNA, we changed to a direct mRNA transfection approach. We made R-luc-4 sites mRNA by in vitro transcription from pRL-TK-4 sites plasmid, and prepared an unrelated firefly luciferase (F-luc) mRNA (Iizuka et al. 1994) as a transfection control. These were transfected into HeLa cells together with miCXCR4. The mRNAs were initially made with a physiological $m^7G(5')ppp(5')G$ cap structure and a poly(A) tail. We used this R-luc-4 sites cap and tail mRNA (Fig. 7.2a) to optimise several transfection parameters for detection sensitivity and extent of miRNA-mediated repression. We settled on transfecting cells with Lipofectamine 2000 (Invitrogen) in 24-well plates, using 20 ng of R-luc test mRNA, 80 ng F-luc control mRNA per well, adding miCXCR4 to a final concentration of 2 nM, and harvesting cells after 16 h of incubation (Clancy et al. 2007). These conditions gave robust and specific miCXCR4-mediated translational repression with very minor effects on mRNA stability (Humphreys et al. 2005). The use of direct mRNA transfection is sometimes criticised because mRNAs introduced into cells in this way have not transited through the nucleus.

Table 7.2 Translation initiation factor requirements of internal ribosome entry site (IRES) elements[a]

IRES:	Mode of initiation:
Hepatitis A Virus (HAV)	All eIFs required
Encephalomyocarditis Virus (EMCV)	Independent of eIF4E
Classical Swine Fever Virus (CSFV) or Hepatitis C Virus (HCV)	Independent of eIF4E, -4G, -4A, -4B
Cricket Paralysis Virus (CrPV), intergenic region	Factor-less initiation from a CGU codon placed in the aminoacyl site of the small ribosomal subunit

[a] This table was previously published in (Clancy et al. 2007)

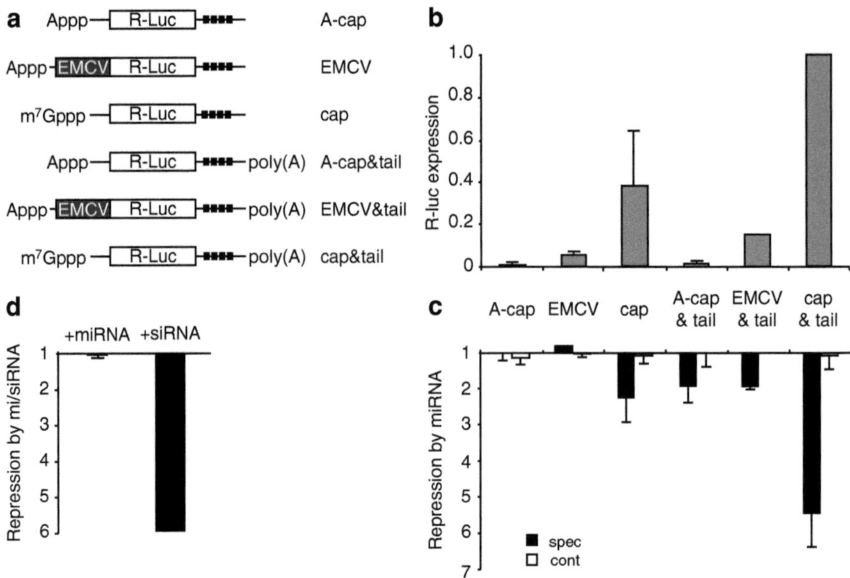

Fig. 7.2 Analysis of miCXCR4 mediated repression of R-luc-4 sites reporter mRNAs after co-transfection into HeLa cells. (**a**) Schematic of the R-luc-4-sites mRNAs used here. (**b**) Normalised R-luc activity of each mRNA transcript from transfections in the absence of miCXCR4. Expression from R-luc-4-sites cap and tail mRNA is set to 1.0. (**c**) Repression of the different R-luc-4-sites mRNAs by the specific miCXCR4 (*filled bars*) or non-specific let-7 miRNA (*open bars*). Average results from three to five experiments are shown with standard deviation. (**d**) Repression of EMCV-R-luc-4-sites by either the miCXCR4 or the perfectly complementary CXCR4 siRNA. Average results from two independent experiments are shown. Illustrations and data contained in this figure were previously published (Humphreys et al. 2005, 2007; Clancy et al. 2007)

We clearly see repression by an miRNA with this approach, indicating that a "nuclear experience" is not required for it. The other common concern is that a proportion of the transfected mRNA may get trapped within inert vesicles (Barreau et al. 2006), complicating the assessment of physical mRNA stability. We routinely address this by additionally measuring the functional half-life of the reporter mRNAs (Gallie 1991), a procedure that is not affected by the above-mentioned issue (Humphreys et al. 2005; Clancy et al. 2007). We further wanted to avoid the complications commonly observed with bi-cistronic IRES reporter approaches (Bert et al. 2006) and instead decided to insert the viral IRES of our choice within the 5′ UTR of mono-cistronic R-luc-4 sites reporter mRNA (Fig. 7.2a). IRES-containing mRNAs were further capped with the non-physiological A(5′)ppp(5′)G cap structure, which is inactive in recruiting the translation initiation machinery but protects the mRNA against accelerated decay (Bergamini et al. 2000). This approach avoids potential problems due to interference of canonical, cap-dependent translation with genuine IRES-mediated translation on the same mRNA template.

7.2.2 *An miRNA Blocks the Initiation Phase of Translation*

The first derivative R-luc-4 sites mRNA we tested contained the IRES from the Cricket Paralysis Virus intergenic region (CrPV IGR), was A-capped and lacked a 3′ poly(A) tail. Translation driven by this IRES retains the canonical aspects of elongation and termination, while initiating from an alanine codon without involvement of Met-tRNA$_i^{met}$ or any eIFs (Hellen and Sarnow 2001; Pestova and Hellen 2003), thus replacing all aspects of regular translation initiation with a radically different mechanism. Translation of this mRNA was inefficient but completely dependent on IRES activity, as shown by comparison to the inactive point mutated CrPV IGRmut14 control (Wilson et al. 2000; Humphreys et al. 2005). Importantly, we found that translation of the CrPV-R-luc-4 sites mRNA was no longer repressible by miCXCR4, indicating that the initiation phase of canonical translation was the major target of miCXCR4-mediated regulation. Other investigators have also used IRES-containing constructs, arriving either at similar or very different conclusions to ours. The reasons for this discrepancy are essentially unknown, as discussed by Valencia-Sanchez et al. (2006) and Jackson and Standart (2007).

Translation initiation on a typical cellular mRNA is jointly promoted by the cap structure and the poly(A) tail. In the process, the mRNA is thought to adopt a closed-loop conformation involving bridging interactions of the adapter protein eIF4G with both, eIF4E bound to the cap structure and the poly(A) binding protein, PABP, bound to the poly(A)-tail (Gallie 1991; Jacobson 1996; Preiss and Hentze 1998, 2003; Amrani et al. 2008). We therefore asked whether miCXCR4 repression targets the function of either the cap structure or the poly(A) tail by preparing four versions of the R-luc-4 sites mRNA: with no poly(A) tail and either an A-cap or a physiological m^{7}G-cap, or with a poly(A) tail and either an A-cap or a physiological m^{7}G-cap (Fig. 7.2a). Measuring the expression of these four R-luc-4 sites mRNA versions in transfected cells reveals the typical functional synergy between the physiological cap and the poly(A) tail: R-luc translation from the mRNA carrying both end modifications was more than the sum of that seen with each individually modified mRNA (Fig. 7.2b). RNA analyses confirmed that all versions of R-luc-4 sites mRNA were similarly stable in transfected cells (Humphreys et al. 2005). Importantly, these mRNAs displayed markedly different responses to miCXCR4 (Fig. 7.2c). The mRNA having neither a physiological cap nor a poly(A) tail ("A-cap"; Fig. 7.3b) was completely resistant to miRNA addition (Fig. 7.3c). Translation driven either solely by the physiological cap ("cap"), or solely by the poly(A) tail ("A-cap and tail"), was only partially responsive to the miCXCR4 (~2-fold repression). A full response to miCXCR4 was only seen with the "cap and tail" version of the mRNA (~5.5-fold repression). The observed effects were target-specific as none of the R-luc-4 sites versions was affected by co-transfection of a control miRNA (*C. elegans let-7*, Fig. 7.2c). These results indicate that the cap structure and poly(A) tail were each necessary but not individually sufficient for full miCXCR4-mediated repression. Our experiments may be criticised for employing a synthetic

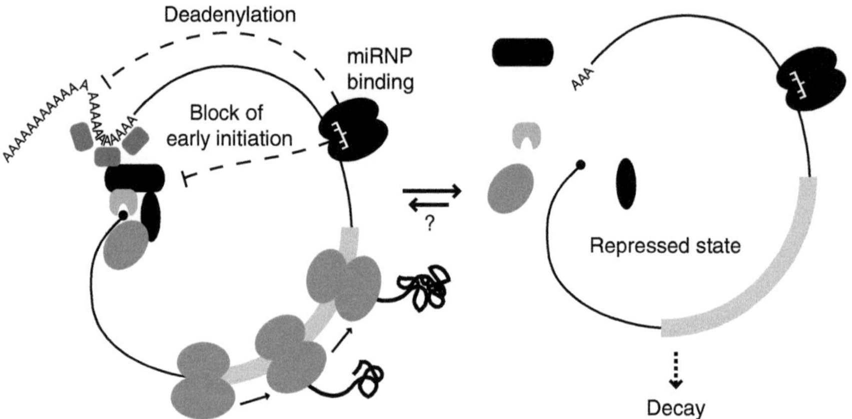

Fig. 7.3 A working model of miRNA-mediated interference of mRNA closed-loop formation during translation initiation. miRNAs guide miRNP complexes to the 3′ UTR of target mRNA. Consequences of miRNP recruitment are to trigger mRNA deadenylation, interference with cap function and possibly aggregation of the mRNA into processing bodies. These changes conspire to attenuate translation of mRNA. The resulting silenced mRNA is either stably stored, or subjected to mRNA decay by decapping and exonucleolytic decay

miRNA; however, work in other systems and in part with endogenous miRNA yielded results compatible with ours (see Table 7.1). We have further seen cap structure as well as poly(A) tail dependence of reporter mRNA repression by endogenous let-7 miRNA (unpublished observation).

To further investigate the target(s) of miRNA-mediated repression within initiation, we created two additional R-luc-4 sites mRNA variants (Fig. 7.2a) carrying the encephalomyocarditis virus (EMCV) IRES (Jackson 2000; Hellen and Sarnow 2001; Poyry et al. 2004). With reference to the closed-loop model, the EMCV IRES is still responsive to poly(A)-mediated stimulation through the eIF4G–PAPBP interaction, but it recruits eIF4G without employing eIF4E (Bergamini et al. 2000; Hellen and Sarnow 2001; Svitkin et al. 2001). Thus, it retains most canonical features of translation initiation but does not require eIF4E (Jackson 2000; Hellen and Sarnow 2001; Poyry et al. 2004). The EMCV IRES-containing R-luc-4 sites mRNA versions were both A-capped, one carried a poly(A) tail ("EMCV and tail") and the other had no tail ("EMCV"; Fig. 7.2a). The EMCV IRES gave robust levels of R-luc expression, which was augmented by the poly(A) tail (Fig. 7.2b). Translation driven solely by the EMCV IRES was found to be completely resistant to miCXCR4 repression (Fig. 7.2c). To exclude that the miCXCR4 target sites are occluded in the EMCV-R-luc-4 sites due to changes in mRNA folding, we transfected a perfectly complementary short interfering CXCR4 RNA variant (siCXCR4) with the mRNA. siCXCR4 led to robust reduction of EMCV-R-luc-4 sites expression (Fig. 7.2d), excluding such a problem. The "EMCV and tail" mRNA produced a partial response to the miCXCR4, comparable in level to the "A-cap and tail" mRNA (Fig. 7.2c), providing further evidence that an miRNA targets the function of the poly(A) tail. Our results with the EMCV-based constructs are consistent with miCXCR4-mediated effects on cap structure and poly(A) tail function during the translation initiation. They further suggest either the recruitment of eIF4E to the cap or its function at the cap, as a target of miRNA action.

7.3 Towards a Model for miRNA-Mediated Translational Control

A number of studies have been published since 2005, whose findings and conclusions substantially overlap with ours (see Table 7.1). Notably, the Filipowicz group, using reporter mRNAs responding to endogenous let-7 miRNA in transfected HeLa cells, also found that the function of the cap structure and eIF4E during translation initiation was targeted, although they did not see a dependence on the poly(A) tail (Pillai et al. 2005). A series of papers have further presented reconstitutions of miRNA-mediated repression in cell-free translation reactions based on either mammalian cell or *Drosophila melanogaster* embryo extracts (Wang et al. 2006; Mathonnet et al. 2007; Thermann and Hentze 2007; Wakiyama et al. 2007). All these studies confirm the importance of the cap structure for repression, while some also report a contribution by the poly(A) tail (Wang et al. 2006; Wakiyama et al. 2007). It has further been shown that miRNAs commonly cause mRNA deadenylation in *D. melanogaster*, zebrafish and mammalian systems (Behm-Ansmant et al. 2006; Giraldez et al. 2006; Wu et al. 2006; Clancy et al. 2007; Wakiyama et al. 2007; Eulalio et al. 2009), often linked to miRNA-stimulated mRNA decay. While deadenylation as a general feature of miRNA-targeted mRNAs is becoming apparent (Eulalio et al. 2009), it is less clear whether deadenylation is a separate mechanism by which miRNA stimulate mRNA decay, or merely an epiphenomenon of translational repression. A third option would be that it is an integral component of miRNA-mediated repression, having downstream effects on both mRNA translation and stability. The latter case would provide an elegant explanation for why we and others have seen a poly(A) tail dependence of miRNA-mediated translational repression (Humphreys et al. 2005; Wang et al. 2006; Wakiyama et al. 2007). Notwithstanding this trend in the literature towards an initation-based mechanism, there have also been several recent publications (Maroney et al. 2006; Nottrott et al. 2006; Petersen et al. 2006; Lytle et al. 2007) reasserting some or all aspects of the original proposal of a post-initiation effect by miRNAs (Olsen and Ambros 1999; Seggerson et al. 2002). Thus, whether or not one mechanism exists or several cannot be resolved at present. Emerging information on dependence of the type of observable mechanism (initiation/post-initiation), or even the directionality of it (repression/activation), on experimental conditions appears to suggest that more than one mechanism can be operational. The choice of transfection method (Lytle et al. 2007), cellular growth state (Bhattacharyya et al. 2006; Vasudevan and Steitz 2007;

Vasudevan et al. 2007) and even reporter gene promoter choice (Kong et al. 2008) have all been flagged as important determinants of the outcome of miRNA action.

Despite the caveats listed above, we believe it is reasonable to construct a working model that is consistent with at least a substantial portion of the published literature. In our model, mRNA closed-loop formation during translation initiation is impaired by an miRNA in at least two ways, by effects on cap/eIF4E function and through mRNA deadenylation (Fig. 7.3). Each mRNA end modification is at the same time important to repression but neither is absolutely required for it, explaining most reports of repression of mRNAs lacking either the cap structure or the poly(A) tail (Humphreys et al. 2005; Pillai et al. 2005; Behm-Ansmant et al. 2006; Giraldez et al. 2006; Wang et al. 2006; Wu et al. 2006; Wakiyama et al. 2007; Eulalio et al. 2008b). The repressed mRNA can then either be stably stored or degraded, the balance of which may be determined by the specific (3′ UTR) sequence context of an mRNA or the given cellular environment, rather than by differences in the initial trigger of repression. This scenario is reminiscent of the maternal mRNA paradigm of poly(A) tail-mediated translational control. Cytoplasmic polyadenylation elements (CPE) in the 3′ UTRs of these mRNAs orchestrate complex patterns of translational repression and activation during oocyte maturation. CPE-mediated regulation involves inhibition of cap function and aggregation of the silenced oligo-adenylated mRNA into cytoplasmic granules similar to the processing bodies linked to the miRNA mechanism. Importantly, CPEs recruit deadenylases and poly(A) polymerases to the mRNA to modulate poly(A) tail length (Hentze et al. 2007; Pique et al. 2008). Intriguingly, miRNA-mediated repression of translation is reportedly also reversible (Bhattacharyya et al. 2006), and capable of activating translation under certain cellular conditions (Vasudevan and Steitz 2007; Vasudevan et al. 2007).

Clearly, this model does not accommodate a post-initiation effect , but is it at least compatible with all reports of initiation-based effects? The answer to this is unfortunately also not in the affirmative. There have now been several reports describing either interactions of the miRNP with, or a dependence of repression on, translation initiation factors that function downstream of closed-loop formation. In one study, eIF6, a factor with 40/60S ribosomal subunit anti-association activity, and proteins of the 60S ribosomal subunits co-purified with a miRNP-related complex, and depletion of eIF6 impaired miRNA function (Chendrimada et al. 2007). Another report, using a mammalian in vitro translation system documented that miRNA-repressed mRNAs purified with 40S but not with 60S ribosomal subunit components (Wang et al. 2008). In *C. elegans*, genetic interactions of let-7 were seen with several translation initiation factors, notably eIF3 (Ding et al. 2008). The description of a cap-binding motif in the Mid domain of Ago proteins and evidence that Ago proteins can compete with eIF4E for cap-binding appeared to have settled the issue (Kiriakidou et al. 2007). However, experiments in *D. melanogaster* cells have renewed doubts, by failing to confirm the relevance of the cap-binding by Ago, or a requirement of eIF6 for repression (Eulalio et al. 2008b). Finally, plausible theoretical arguments have been advanced, suggesting that several experiments underpinning the proposal of a direct cap/eIF4E targeting by miRNAs could also be explained by a later step in initiation being affected, assuming certain kinetic relationships between different sub-steps of initiation (Nissan and Parker 2008).

7.4 Concluding Remarks

How miRNAs affect translation clearly remains a contentious issue, with several seemingly incompatible models having been advanced. Our own work supports miRNA effects on early roles of the cap structure and poly(A) tail during translation initiation, analogous to several more established paradigms of translational control. Progress in this protracted research field could come from several directions. A careful reproduction of some of the key experiments that gave rise to the divergent models should be tackled, perhaps with an emphasis on the intriguing but also disconcerting dependence of observed effects on experimental conditions. It seems intuitive to posit that miRNAs should act through some form of universal, if multi-faceted, mechanism and the search for such a unifying principle should continue. Given the prominence given to a role of the mRNA cap structure by several studies, it will be particularly important to identify, beyond doubt, the molecular components that bridge between the miRNP and the cap. mRNA deadenylation also warrants further study as another candidate for a proximal, and perhaps universal, miRNA effect, with roles in both translation and decay. Besides its role in early translation initiation, the poly(A) tail can also stimulate the 60S subunit joining step during initiation and it is linked to translation termination, perhaps providing ways to rationalise a role for deadenylation in several modes of translational repression by miRNAs.

Acknowledgements The authors wish to thank Jennifer L Clancy, Belinda J Ryan (née Westman), Marco Nousch, and David IK Martin, who have all contributed to the research discussed in this chapter. Research in the authors' lab was supported by grants from the National Health and Medical Research Council, the Australian Research Council, the Sylvia and Charles Viertel Charitable Foundation, and by the Victor Chang Cardiac Research Institute.

Note Added in Proof

Since the above was written, new work in Drosophila extracts and transfected mammalian cells has directly shown that a miRNP primarily targets the mRNA cap structure and that deadenylation can augment miRNA-mediated translational repression (Beilharz et al., 2009; Zdanowicz et al., 2009). Work in a mammalian in vitro translation system further revealed that the miRNP interacts with PABP and recruits the CAF1/CCR4 deadenylase complex to the mRNA (Fabian et al., 2009).

References

Amaral PP, Dinger ME, Mercer TR, Mattick JS (2008) The eukaryotic genome as an RNA machine. Science 319(5871):1787–1789

Ambros V (2004) The functions of animal microRNAs. Nature 431(7006):350–355

Amrani N, Ghosh S, Mangus DA, Jacobson A (2008) Translation factors promote the formation of two states of the closed-loop mRNP. Nature 453(7199):1276–1280

Barreau C, Dutertre S, Paillard L, Osborne HB (2006) Liposome-mediated RNA transfection should be used with caution. RNA 12(10):1790–1793

Bartel DP, Chen CZ (2004) Micromanagers of gene expression: the potentially widespread influence of metazoan microRNAs. Nat Rev Genet 5(5):396–400

Behm-Ansmant I, Rehwinkel J, Doerks T, Stark A, Bork P, Izaurralde E (2006) mRNA degradation by miRNAs and GW182 requires both CCR4:NOT deadenylase and DCP1:DCP2 decapping complexes. Genes Dev 20(14):1885–1898

Beilharz TH, Humphreys DT, Clancy JL, Thermann R, Martin DIK, Hentze MW, Preiss T (2009) microRNA-mediated messenger RNA deadenylation contributes to translational repression in mammalian cells. PLoS ONE 4, e6783

Bergamini G, Preiss T, Hentze MW (2000) Picornavirus IRESes and the poly(A) tail jointly promote cap- independent translation in a mammalian cell-free system. RNA 6(12):1781–1790

Bert AG, Grepin R, Vadas MA, Goodall GJ (2006) Assessing IRES activity in the HIF-1alpha and other cellular 5’ UTRs. RNA 12(6):1074–1083

Bhattacharyya SN, Habermacher R, Martine U, Closs EI, Filipowicz W (2006) Relief of microRNA-mediated translational repression in human cells subjected to stress. Cell 125(6):1111–1124

Borman AM, Kean KM (1997) Intact eukaryotic initiation factor 4G is required for hepatitis A virus internal initiation of translation. Virology 237:129–136

Carthew RW (2006) Gene regulation by microRNAs. Curr Opin Genet Dev 16(2):203–208

Chang TC, Mendell JT (2007) microRNAs in vertebrate physiology and human disease. Annu Rev Genom Hum Genet 8:215–239

Chendrimada TP, Finn KJ, Ji X, Baillat D, Gregory RI, Liebhaber SA, Pasquinelli AE, Shiekhattar R (2007) MicroRNA silencing through RISC recruitment of eIF6. Nature 447(7146):823–828

Clancy JL, Nousch M, Humphreys DT, Westman BJ, Beilharz TH, Preiss T (2007) Methods to analyze microRNA-mediated control of mRNA translation. Methods Enzymol 431:83–111

Ding XC, Slack FJ, Grosshans H (2008) The let-7 microRNA interfaces extensively with the translation machinery to regulate cell differentiation. Cell Cycle 7(19):3083–3090

Doench JG, Sharp PA (2004) Specificity of microRNA target selection in translational repression. Genes Dev 18(5):504–511

Doench JG, Petersen CP, Sharp PA (2003) siRNAs can function as miRNAs. Genes Dev 17(4):438–442

Eulalio A, Behm-Ansmant I, Izaurralde E (2007) P bodies: at the crossroads of post-transcriptional pathways. Nat Rev Mol Cell Biol 8(1):9–22

Eulalio A, Huntzinger E, Izaurralde E (2008a) Getting to the root of miRNA-mediated gene silencing. Cell 132(1):9–14

Eulalio A, Huntzinger E, Izaurralde E (2008b) GW182 interaction with Argonaute is essential for miRNA-mediated translational repression and mRNA decay. Nat Struct Mol Biol 15(4):346–353

Eulalio A, Huntzinger E, Nishihara T, Rehwinkel J, Fauser M, Izaurralde E (2009) Deadenylation is a widespread effect of miRNA regulation. RNA 15(1):21–32

Fabian MR, Mathonnet G, Sundermeier T, Mathys H, Zipprich JT, Svitkin Y, Rivas F, Jinek M, Wohlschlegel J, Doudna JA, Chen CYA, Shyu AB, Yates III JR, Hannon GJ, Filipowicz W, Duchaine TF, Sonenberg N (2009) Mammalian miRNA RISC recruits CAF-1 and PABP to affect PABP-dependent deadenylation. Mol Cell doi:10.1016/j.molcel.2009.08.004

Faller M, Guo F (2008) MicroRNA biogenesis: there’s more than one way to skin a cat. Biochim Biophys Acta 1779(11):663–667

Filipowicz W, Bhattacharyya SN, Sonenberg N (2008) Mechanisms of post-transcriptional regulation by microRNAs: are the answers in sight? Nat Rev Genet 9(2):102–114

Fraser CS, Doudna JA (2007) Structural and mechanistic insights into hepatitis C viral translation initiation. Nat Rev Microbiol 5(1):29–38

Friedman RC, Farh KK, Burge CB, Bartel DP (2009) Most mammalian mRNAs are conserved targets of microRNAs. Genome Res 19(1):92–105

Gallie DR (1991) The cap and poly(A) tail function synergistically to regulate mRNA translational efficiency. Genes Dev 5(11):2108–2116

Gebauer F, Hentze MW (2004) Molecular mechanisms of translational control. Nat Rev Mol Cell Biol 5(10):827–835

Giraldez AJ, Mishima Y, Rihel J, Grocock RJ, Van Dongen S, Inoue K, Enright AJ, Schier AF (2006) Zebrafish MiR-430 Promotes Deadenylation and Clearance of Maternal mRNAs. Science 312(5770):75–79

Hellen CU, Sarnow P (2001) Internal ribosome entry sites in eukaryotic mRNA molecules. Genes Dev 15(13):1593–1612

Hentze MW, Gebauer F, Preiss T (2007) Cis-regulatory sequences and trans-acting factors in translational control. In Mathews MB, Sonenberg N, Hershey JWB (eds) Translational control in biology and medicine. Cold Spring Harbor Laboratory, Cold Spring Harbor, New York, pp 269–295

Humphreys DT, Westman BJ, Martin DI, Preiss T (2005) MicroRNAs control translation initiation by inhibiting eukaryotic initiation factor 4E/cap and poly(A) tail function. Proc Natl Acad Sci U S A 102(47):16961–16966

Humphreys DT, Westman BJ, Martin DI, Preiss T (2007) Inhibition of translation initiation by a microRNA. In: Appasani K (ed) MicroRNAs: from basic science to disease biology. Cambridge University Press, Cambridge, UK, pp 85–101

Iizuka N, Najita L, Franzusoff A, Sarnow P (1994) Cap-dependent and cap-independent translation by internal initiation of mRNAs in cell extracts prepared from Saccharomyces cerevisiae. Mol Cell Biol 14(11):7322–7330

Jackson RJ (2000) Comparative view of initiation site selection mechanisms. In Sonenberg N, Hershey JBW, Mathews MB (eds) Translational control of gene expression. Cold Spring Harbor Laboratory, Cold Spring Harbor, New York, pp 127–184

Jackson RJ (2005) Alternative mechanisms of initiating translation of mammalian mRNAs. Biochem Soc Trans 33(Pt 6):1231–1241

Jackson RJ, Standart N (2007) How do microRNAs regulate gene expression? Sci STKE 2007(367):re1

Jacobson A (1996) Poly(A) metabolism and translation: the closed-loop model. In Hershey JWB, Mathews MB, Sonenberg N (eds) Translational control. Cold Spring Harbor Laboratory, Cold Spring Harbor, New York, pp 451–480

Kapp LD, Lorsch JR (2004) The molecular mechanics of eukaryotic translation. Annu Rev Biochem 73:657–704

Kiriakidou M, Tan GS, Lamprinaki S, De Planell-Saguer M, Nelson PT, Mourelatos Z (2007) An mRNA m7G cap binding-like motif within human Ago2 represses translation. Cell 129(6):1141–1151

Kloosterman WP, Plasterk RH (2006) The diverse functions of microRNAs in animal development and disease. Dev Cell 11(4):441–450

Kong YW, Cannell IG, de Moor CH, Hill K, Garside PG, Hamilton TL, Meijer HA, Dobbyn HC, Stoneley M, Spriggs KA, Willis AE, Bushell M (2008) The mechanism of micro-RNA-mediated translation repression is determined by the promoter of the target gene. Proc Natl Acad Sci U S A 105(26):8866–8871

Lytle JR, Yario TA, Steitz JA (2007) Target mRNAs are repressed as efficiently by microRNA-binding sites in the 5' UTR as in the 3' UTR. Proc Natl Acad Sci U S A 104(23):9667–9672

Maroney PA, Yu Y, Fisher J, Nilsen TW (2006) Evidence that microRNAs are associated with translating messenger RNAs in human cells. Nat Struct Mol Biol 13(12):1102–1107

Mathonnet G, Fabian MR, Svitkin YV, Parsyan A, Huck L, Murata T, Biffo S, Merrick WC, Darzynkiewicz E, Pillai RS, Filipowicz W, Duchaine TF, Sonenberg N (2007) MicroRNA inhibition of translation initiation in vitro by targeting the cap-binding complex eIF4F. Science 317(5845):1764–1767

Nissan T, Parker R (2008) Computational analysis of miRNA-mediated repression of translation: implications for models of translation initiation inhibition. RNA 14(8):1480–1491

Nottrott S, Simard MJ, Richter JD (2006) Human let-7a miRNA blocks protein production on actively translating polyribosomes. Nat Struct Mol Biol 13(12):1108–1114

Olsen PH, Ambros V (1999) The lin-4 regulatory RNA controls developmental timing in Caenorhabditis elegans by blocking LIN-14 protein synthesis after the initiation of translation. Dev Biol 216(2):671–680

Ostareck DH, Ostareck-Lederer A, Shatsky IN, Hentze MW (2001) Lipoxygenase mRNA silencing in erythroid differentiation: the 3'UTR regulatory complex controls 60S ribosomal subunit joining. Cell 104(2):281–290

Pestova TV, Hellen CU (2003) Translation elongation after assembly of ribosomes on the Cricket paralysis virus internal ribosomal entry site without initiation factors or initiator tRNA. Genes Dev 17(2):181–186

Peters L, Meister G (2007) Argonaute proteins: mediators of RNA silencing. Mol Cell 26(5):611–623

Petersen CP, Bordeleau ME, Pelletier J, Sharp PA (2006) Short RNAs repress translation after initiation in mammalian cells. Mol Cell 21(4):533–542

Pillai RS, Bhattacharyya SN, Artus CG, Zoller T, Cougot N, Basyuk E, Bertrand E, Filipowicz W (2005) Inhibition of translational initiation by let-7 MicroRNA in human cells. Science 309(5740):1573–1576

Pique M, Lopez JM, Foissac S, Guigo R, Mendez R (2008) A combinatorial code for CPE-mediated translational control. Cell 132(3):434–448

Poyry TA, Kaminski A, Jackson RJ (2004) What determines whether mammalian ribosomes resume scanning after translation of a short upstream open reading frame? Genes Dev 18(1):62–75

Preiss T, Hentze MW (1998) Dual function of the messenger RNA cap structure in poly(A)-tail-promoted translation in yeast. Nature 392:516–520

Preiss T, Hentze MW (2003) Starting the protein synthesis machine: eukaryotic translation initiation. Bioessays 25(12):1201–1211

Seggerson K, Tang L, Moss EG (2002) Two genetic circuits repress the Caenorhabditis elegans heterochronic gene lin-28 after translation initiation. Dev Biol 243(2):215–225

Sonenberg N, Dever TE (2003) Eukaryotic translation initiation factors and regulators. Curr Opin Struct Biol 13(1):56–63

Standart N, Jackson RJ (2007) MicroRNAs repress translation of m7G ppp-capped target mRNAs in vitro by inhibiting initiation and promoting deadenylation. Genes Dev 21(16):1975–1982

Svitkin YV, Imataka H, Khaleghpour K, Kahvejian A, Liebig HD, Sonenberg N (2001) Poly(A)-binding protein interaction with eIF4G stimulates picornavirus IRES-dependent translation. RNA 7(12):1743–1752

Thermann R, Hentze MW (2007) Drosophila miR2 induces pseudo-polysomes and inhibits translation initiation. Nature 447(7146):875–878

Valencia-Sanchez MA, Liu J, Hannon GJ, Parker R (2006) Control of translation and mRNA degradation by miRNAs and siRNAs. Genes Dev 20(5):515–524

Vasudevan S, Steitz JA (2007) AU-rich-element-mediated upregulation of translation by FXR1 and Argonaute 2. Cell 128(6):1105–1118

Vasudevan S, Tong Y, Steitz JA (2007) Switching from repression to activation: microRNAs can up-regulate translation. Science 318(5858):1931–1934

Wakiyama M, Takimoto K, Ohara O, Yokoyama S (2007) Let-7 microRNA-mediated mRNA deadenylation and translational repression in a mammalian cell-free system. Genes Dev 21(15):1857–1862

Wang B, Love TM, Call ME, Doench JG, Novina CD (2006) Recapitulation of short RNA-directed translational gene silencing in vitro. Mol Cell 22(4):553–560

Wang B, Yanez A, Novina CD (2008) MicroRNA-repressed mRNAs contain 40S but not 60S components. Proc Natl Acad Sci U S A 105(14):5343–5348

Wilson JE, Powell MJ, Hoover SE, Sarnow P (2000) Naturally occurring dicistronic cricket paralysis virus RNA is regulated by two internal ribosome entry sites. Mol Cell Biol 20(14):4990–4999

Wu L, Belasco JG (2008) Let me count the ways: mechanisms of gene regulation by miRNAs and siRNAs. Mol Cell 29(1):1–7

Wu L, Fan J, Belasco JG (2006) MicroRNAs direct rapid deadenylation of mRNA. Proc Natl Acad Sci U S A 103(11):4034–4039

Zdanowicz A, Thermann R, Kowalska J, Jemielity J, Duncan K, Preiss T, Darzynkiewicz E, Hentze MW (2009) Drosophila miR2 primarily targets the m7GpppN cap structure for translational repression. Mol Cell in press.